Artillery Tactics of the Civil War

A Study of the Tactical Use of Artillery Based on the First Day's Battle at Gettysburg

by

R.L. Murray

Benedum Books
13205 Younglove Road
Wolcott, NY 14590

www.nyincivilwar.com

I.S.B.N.

Paperback 0-9646261-6-0

Revised Edition

This book is dedicated to my wonderful children:

Ryan

Patrick

Rebekah

Katherine

God bless you.

Contents

Acknowledgments

This book originated as a project for my Master's program. I would like to acknowledge and thank an excellent teacher and historian, Dr. Charles White, my professor during the project. His encouragement and insight were very valuable.

A special thanks to Ron Patrick and George W. Contant for editing and writing suggestions. Their assistance was greatly appreciated.

I would also like to thank: Sean McAdoo, Connie Benowitz, Mary Ruth Hughes, Russell C. Downes, and Todd Shillington.

Introduction

At the beginning of the American Civil War the artillery arm had a very good reputation. During the war with Mexico the United States artillery distinguished itself on several fields of battle, proving to be among the best in the world.[1] Thus, when the war between the states began, many believed the field pieces would play an even more prominent role than ever before.[2]

During the first two years of the Civil War artillery did play a significant role, but often times terrain and inefficient organization limited its effectiveness. Unlike most of the battlefields in Mexico, the forested regions of northern Virginia – which played host to many of the major conflicts in the Eastern theater during the Civil War – restricted the visibility and the field of fire for the artillerymen, thus limiting their potential.

Even with the wooded terrain, there were several times during the first two years of the war when the artillery did distinguish itself. At Malvern Hill the Federals used their massed batteries to crush an assault by the Southern infantry. The Confederates in turn used their guns to break up the attacking Union lines at Second Manassas, Antietam and Fredericksburg.[3] The initial seasons of campaigning were times of learning for both armies. By 1863, there were both seasoned cannoneers to man the batteries, and new organizational structures for the artillery that greatly increased its effectiveness (discussed in detail later).

In many ways 1863 was the climax year for the use of field artillery in the Eastern theater, and the Battle of Gettysburg

---

[1] John S.D. Eisenhower, *So Far From God: The U.S. War with Mexico 1846 – 1848* (New York: 1989), xxii – xxiii and 379 – 380.

[2] Grady McWhiney and Perry Jamieson, *Attack and Die: Civil War Military Tactics and the Southern Heritage* (Tuscalusa, 1982), 112. "Winfield Scott believed that the artillery would be the most important arm in the Civil War."

[3] *Ibid.*, 112 – 113.

possibly was the zenith for the entire war.[4] At Gettysburg both sides were able to utilize effective organizational arrangements for their batteries. These new systems, combined with the relatively open countryside at Gettysburg, allowed the artillerists to demonstrate their impressive tactical skills.

---

[4] During the campaigns of 1864 (in the East), the artillery's effectiveness was again greatly limited by the wooded terrain. General Grant even made the decision to limit the number of his guns accompanying his army. "Finding the country through which it [the Army of the Potomac] had to move was more difficult than he had supposed, General Grant gave an order on the 19th to send ninety-two guns back to Washington." General Horace Porter, *Campaigning with Grant* (New York, 1897), 129.

Chapter One

Civil War Artillery

There were many types of cannon used during the Civil War, but the most common on the battlefields in the East were the artillery guns that fired ten to twelve pound projectiles. These "light" field pieces were divided into two basic classifications: the smoothbore and the rifle.

Unlike the muskets carried by the infantry, the rifled cannon did not have a dramatic advantage over their smoothbore counterparts.[1] Rifling technology quickly made smoothbore muskets obsolete. Because the smoothbores lacked the accuracy and range of rifled muskets, their relative value decreased significantly. Interestingly, the smoothbore cannon also lacked the range and accuracy of their rifled counterparts, but their relative value remained high.

One of the reasons smoothbore cannon retained their value was that Civil War artillery's most effective use was in firing canister, and smoothbores were more deadly than rifled pieces when discharging this type of ordnance.[2] Canister rounds were tin cans filled with small iron or lead balls packed in sawdust. The force of

---

[1] McWhiney and Jamieson, *Attack and Die,* 122.

[2] L. Van Loan Naisawald, *Grape and Canister: The Story of Field Artillery in the Army of the Potomac* (New York, 1960), 539. "The only really effective round of the Civil War was canister."; Edwin B. Coddington, *The Gettysburg Campaign: A Study in Command* (New York, 1968), 251. "No other gun could match it [the smoothbore12-pound Napoleon] for effectiveness against infantry at close range."; Paddy Griffith, *Battle Tactics of the Civil War* (New Haven, 1987), 170.

the discharge caused the balls to spray out of the muzzle, turning the cannon into a "huge shotgun."[3] With a larger diameter bore (4.62-inches compared to 3 or 2.9 inches of the rifled pieces), the smoothbore 12-pound cannon could discharge more projectiles and in a wider pattern than rifled pieces.[4]

Aside from canister, there were other advantages associated with smoothbore artillery. Rifled guns used cone shaped ordnance that rotated or spiraled in flight. While the shape and rotation allowed the projectiles to travel further and in a more accurate line, oftentimes their effectiveness on impact was not as great. The main problem was that when the rotating projectiles struck the ground, they might 'corkscrew' deep into the dirt. Thus, the same properties that increased accuracy in flight could also reduce effectiveness when the projectiles reached their target.[5]

Conversely, the round shaped ordnance fired from smoothbore cannon often bounded along on the hard ground after impact. This increased the probability of hitting the enemy; a projectile that fell short of its mark might well continue along until it reached its target (or another). In Napoleonic times this was known as "ricochet fire." Gunners actually aimed short of the target, planning for the projectiles to skip or bound on to the objective. This tactic often proved more effective than aiming directly

---

[3] Naisawald, *Grape and Canister,* 539.

[4] James C. Hazlett, Edwin Olmstead, and M. Hume Parks, *Field Artillery Weapons of the Civil War* Newark, N.J., 1988), 88: Naisawald, *Grape and Canister,* 551. The rifling in the barrel also slightly decreased the effectiveness of canister fired from a rifled piece. The spiral groves affected the flight of some of the small projectiles as they traveled down the barrel.

[5] R.L. Murray, *"Hurrah for the Ould Flag!: The True Story of Captain Andrew Cowan and the First New York Independent Battery at Gettysburg* (Wolcott, NY, 1998), 50. A member of the battery described the fear of an incoming shell striking the ground in front of him, but to his relief it was fired from a rifled gun and "buried itself" into the sod. He escaped unharmed; Harry W. Pfanz, *Gettysburg: Culp's Hill and Cemetery Hill* (Chapel Hill, North Carolina, 1993), 181. During the artillery duel on East Cemetery Hill on July 2, Col. Wainwright watched as a shell from a rifled piece struck the ground in front of him and burrowed itself into the dirt. The shell exploded deep in the ground beneath him and he was not injured.

at the targets.[6]

These factors served to keep the smoothbores in service, but in many ways they could not match the rifled cannon. Rifling gave the field pieces a decided advantage in long range counterbattery fire. By staying outside of the 1200 yard effective range of the smoothbore Napoleons, the rifled guns could accurately fire shot and shell at the enemy and usually inflict greater damage than they received.[7] It is important to recognize that each type of gun had its own merits. Smoothbores were most effective at close range, especially against massed infantry, while the rifled pieces had a longer range and greater accuracy for striking enemy batteries or distant troops.

The basic operational command for artillery was the battery. While size and gun uniformity differed in the opposing armies, batteries generally contained four to six field pieces and were manned with approximately 80 to 150 soldiers. At Gettysburg most Federal batteries contained six guns, while the Confederate units usually had four pieces. Other than size, the Federals also had an advantage in greater uniformity – containing the same type of guns, such as six twelve-pound Napoleons or six three-inch ordinance rifles.[8] The mixing of cannon often created ordnance problems. Because the limbers and caissons had to carry rounds for their respective pieces rather than for the entire command, it was easy for the crewmen to grab the wrong type of ammunition

---

[6] Brent Nosworthy, *With Musket Cannon and Sword: Battle Tactics of Napoleon and His Enemies* (New York, 1996), 387 – 391.

[7] Coddington, *The Gettysburg Campaign,* 251; Naisawald, *Grape and Canister,* 36.

[8] Jennings Cropper Wise, *The Long Arm of Lee,* or *The History of the Artillery of the Army of Northern Virginia* (Lynchburg, Va., 1915), 571. Citied hereinafter as Wise, *The Long Arm of Lee*; John W. Busey and David Martin, *Regimental Strengths and Losses at Gettysburg* (Hightstown, N.J.,1982). Busey and Martin list the makeup of guns in each battery, Union and Confederate. Examining this list one soon notices that the Confederate batteries were a mixture of pieces, while the Federal batteries were almost without exception made up of the same type of cannon.

during a battle.[9]

Within the battery there were smaller commands known as sections – each containing two artillery pieces. Sections could have been detached to fight separately, or assigned different targets (one section focusing on a threat to the battery's flank while the others dueled with guns on their front). The tactics employed by the artillery officer often included decisions involving the deployment (or choice of targets) for his individual sections.

On the battlefield the battery commander had many difficult decisions to make: where and how to deploy, what target to choose (infantry or the enemy's artillery), where to place his caissons for re-supply, and, if necessary, how best to disengage and retire. When given the opportunity to operate free of interference (mainly from nearby infantry commanders – more on this in the next chapter), the battery commander usually tried to mass his guns and choose a single target to fire upon. He tried to deploy the cannon where they had a good view of the enemy (such as a slightly elevated piece of ground), while avoiding unnecessary exposure to enemy fire. The difficulty was in maximizing the cannon's effectiveness, while at the same time minimizing the risks to his guns and crew.

Artillerists also tried to deploy their guns on smooth level terrain. This type of surface allowed the crews to work more efficiently and made the sighting of the piece easier (an adjustment in the trajectory was needed if the cannon was on uneven ground). Additionally, because each discharge caused the gun to change position, the crews needed to continually re-aim the piece accord-

[9] Wise, *The Long Arm of Lee*, 571. "...a distressing lack of uniformity in material existed. This was of course a glaring defect, greatly increasing the difficulty of ammunition supply and impairing the general efficiency." Several factors worked against the logical decision to incorporate similar weapons into batteries. One of the largest problems was that the Confederates still had many of the obsolete howitzers in service, and no one wanted a battery made up entirely of these guns. Additionally, the men were used to their particular weapons and were not willing to trade them for strange guns.

ing to the gunner's direction.[10] Rough or uneven ground made this task more difficult.

Another important consideration was the angle of fire on the enemy lines. Always desiring to maximize the effect of each discharge, cannoneers looked for opportunities to strike the enemy's flank. Rather than firing perpendicularly at a target, they tried to increase the destruction by firing at an obtuse angle. This was known as enfilade fire.

Illustration 3-1: Enfilade Fire

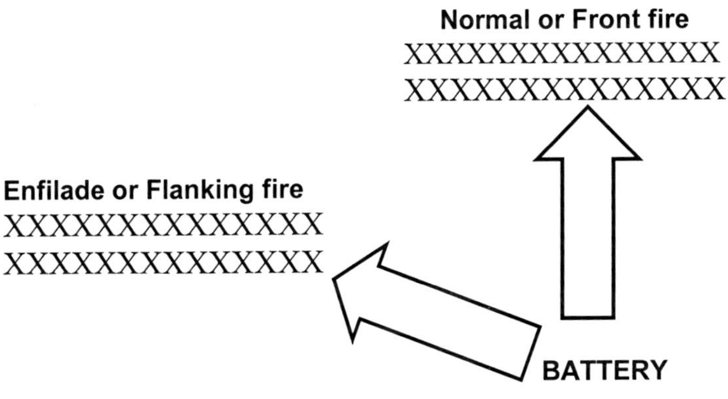

Enfilade fire was more destructive for two reasons. First, the dimension of depth was greater; shots aimed too high would still have a chance of hitting troops or guns in line beyond the target. Second, once shots found their mark, they usually inflicted more damage upon the enemy. Rather than passing through two ranks of infantry, the projectile might continue down the line causing more destruction. Flanking (or enfilade) fire was an important factor for the artillery officer to consider, both trying to achieve it and avoid

[10] Harry W. Pfanz, *Gettysburg: The Second Day* (Chapel Hill, 1987), 125.

falling victim to it.

The largest operational unit for artillery was either the battalion or brigade (depending on whether it was a Confederate or Federal unit – the details of each will be discussed in the second chapter). Organizing the artillery into these larger units allowed for more effective utilization of the army's cannon. The goal was to increase both the number of guns engaged and the coordination of fire between individual batteries. Artillery was most effective when used in a concentrated and coordinated effort, compared to uncoordinated piecemeal deployments.[11]

Tactical considerations for the battalion or brigade were much the same as with a single battery, just on a larger scale. Ideally, the commander looked for an elevated position that offered some shelter, and tried to enfilade the enemy's position whenever possible.

---

[11] Griffith, *Battle Tactics of the Civil War,* 166.

Chapter Two

The Reorganization of Artillery

Napoleon once said, "Artillery should always be placed in the most advantageous positions...."[1] The difficult task for the Civil War commander was deciding where on the field the "most advantageous positions" were. Two general schools of thought concerning artillery deployment existed at the beginning of the Civil War – each with its own merits. Supporters of one theory believed it was essential for artillery pieces to be spread throughout the ranks; lines held by raw recruits needed artillery support to insure proper defensive strength. The other theory called for the massing of guns, allowing for concentration of fire at decisive points on the field. Concentration, they maintained, increased the effectiveness of artillery and thus was the key to its deployment.

With the expansion of the armies in late 1861 and early 1862, the former theory was implemented. Most commanders believed that the dispersion of field pieces was essential in an army of inexperienced soldiers. Few discounted the value of the massing of guns, but the necessity of maintaining the line's defensiveness took priority.

The inefficiency associated with the dispersion of guns became readily apparent during the initial campaigns. While the bat-

---

[1] David Chandler, ed., *The Military Maxims of Napoleon* (Stackpole Books edition, 1994), 73.

tle raged on one part of the field, a large number of cannon remained idle on another. Commanders hesitated to send "their" artillery to assist in other locations, fearing the possibility of an attack on their own front.[2] The commanding generals became frustrated with this system. They recognized that better utilization of the artillery might make the difference in future battles.

Another problem leading to inefficient deployment involved the ranks held by artillery officers. Lower ranking commissions meant that skilled battery commanders were often overruled in selecting locations and targets for their guns. Infantry or cavalry officers in command of neighboring units, whose expertise in artillery matters was often lacking, could dictate gun positioning because of their superior rank. This led to inefficiency and frustration for the cannoneers.

Promotions for artillery officers were very slow compared to those in the infantry and cavalry. Additionally, until the battalions and brigade systems were created, battery commanders had no real hope of promotion beyond captain (with few exceptions). This forced many officers to leave their artillery commands to gain deserved promotions.[3]

---

[2] Stephen W. Sears, *Chancellorsville* (Boston, 1996), 51. "In the crisis of battle the guns were not always where they were most needed, or not in sufficient numbers, or not on hand at the right moment." Douglas Southall Freeman also noted that in the Confederate army the artillery's attachment to a particular brigade made it tough to send it off to serve with another. Douglas Southall Freeman, *Lee's Lieutenants: A Study in Command*, (New York, 1943), Vol. 2, 447. Freeman wrote, "Soon the batteries had become so habituated to serve with particular Brigades that any attempt to send the guns elsewhere was regarded as the darkest injustice to the gunners and to the Brigade..."

[3] Naisawald, *Grape and Canister*; Wise, *The Long Arm of Lee*, 572 -574; Griffith, *Battle Tactics*, 165. Stephen H. Weed is an excellent example of the loss of a talented artillery officer due to the lack of promotion. This West Point graduate served as an artillery officer in the Army of the Potomac for two years without advancing in rank. In fact, even when he was given the responsibility of commanding the Fifth Corps artillery he still remained a captain. Frustrated, he transferred to the infantry where he was given the command of a brigade in June of 1863 - and jumped in rank from captain to brigadier general.

As the war progressed, both armies gradually began reorganizing their artillery into larger commands. While the goal of these changes was to increase the firepower and flexibility of the field pieces, it also provided promotion opportunities for the artillerists (especially for the Confederates – see Appendix Two for a list of artillery commanders and their ranks at Gettysburg). The potential for promotions helped encourage talented officers into remaining with their artillery units.

By the autumn of 1862, the Confederates placed most of the batteries in the Army of Northern Virginia under divisional command – rather than having them assigned to individual brigades. While this was an improvement, there were still problems that inhibited efficient concentrations of firepower. The main obstacle was the reluctance of divisional commanders to send their guns off to another part of the field. These officers wanted to keep their artillery close at hand, much the same as the brigade commanders had done. This understandable desire to keep control of one's own batteries hurt the army's overall firepower.[4]

Lee's Chief of Artillery, Brigadier General William Nelson Pendleton, moved to correct this problem after the battle of Fredericksburg. "Batteries... can scarcely be assigned elsewhere, whatever the emergency," Pendleton wrote to General Lee, "without producing some difficulty, almost as if a vested right was violated. But most injurously [sic] of all, this system hinders unity and concentration in battle."[5] To overcome these problems Pendleton recommended that the Army of Northern Virginia adopt a battalion system. Lee and Pendleton discussed the plan and on

---

Ironically, he was killed at Gettysburg on July 2, where he fell a few yards from one of the batteries he used to command. Mark Boatner, *The Civil War Dictionary* (New York, 1959); Stewart Sifakis, *Who Was Who in the Civil War* (New York, 1988); Harry W. Pfanz, *Gettysburg: The Second Day*.

[4] Freeman, *Lee's Lieutenants*, Vol. 2, 447.

[5] *The War of the Rebellion: A Compilation of the Official Records of the Union and Confederate Armies* (Washington: Government Printing Office, 1889), Series I, Serial No. 44, 1030, Vol. 25, Pt. II, 614. Cited hereinafter as *O.R.*, all references being to Series I.

February 15, 1863, General Lee issued General Orders No. 20, reorganizing the artillery.[6]

Both men felt that in order for the Confederate artillery to compete with the more numerous guns in the Army of the Potomac, they needed to be more efficiently organized. To do this they grouped the batteries into battalions, most consisting of sixteen guns – four batteries containing four guns each.[7] These consolidated battery groups were then reassigned at the divisional and corps level. One battalion was attached to each division, and two additional commands were designated as the corps reserve. The remaining two battalions created the army's Artillery Reserve. This gave the artillerists more latitude in deployment because it reduced the local infantry commander's ability to interfere with their guns.[8]

The new battalions were placed under the control of the corps chief of artillery. Overall supervision of the battalions belonged to the chief of artillery for the army, General Pendleton (who reported directly to General Lee). This increased the army's flexibility and firepower for utilizing their field guns; now it would be possible to quickly mass cannon at a strategic point on the field.[9]

---

[6] *O.R.*, Vol. 25, Pt. II, 625 and 728 - 730. The final orders and details were completed in Lee's Special Orders No. 106, issued on April 16, 1863.

[7] *O.R.*, Vol. 25, Pt. II, 625 and 728 - 730.: Ernest B. Furgurson, Chancellorsville 1863: The Souls of the Brave (New York, 1992), 49.

[8] *O.R.*, Vol. 25, Pt. II, 625 – 626; Sears, *Chancellorsville,* 52. Sears states, "The primary gain of the new system was to give more tactical control to the artillerists themselves...control of the guns in battle was shifting to the men most expert in how to exercise it."; Wise, *Long Arm of Lee,* 278. "It was now generally perceived...that the old system of isolating batteries by attaching them to brigades simply crippled the power of concentrating them in action."

[9] Sears, *Chancellorsville,* 52; Furgurson, *Chancellorsville 1863,* 49. Another advantage was in supplying the batteries. General Pendleton noted the weakness in the previous system, where the brigade and divisional supply...officers...cannot devote to one or more batteries the time and attention they imperatively need."

The new battalions would be provided with their own staff for assuring that the batteries were properly supplied rather than relying on an infantry commander to appropriate their special needs. *O.R.*, Vol. 25, Pt. II, 614.

Brigadier General William N. Pendleton

Photograph from Miller's Photographic History of the Civil War

Brigadier General Henry J. Hunt

Photo courtesy of the Gettysburg National Military Park

The new system was to prove itself many times throughout the remainder of the war, including at its first trial in May of 1863.

During the Chancellorsville campaign, General Lee's forces were severely outnumbered (they went into battle minus two of Longstreet's divisions).[10] Fortunately for Lee his new artillery organization performed superbly, helping to make up for his army's deficiency in numbers. Because of the battalion system (and the mistakes that Major General Joseph Hooker made with his artillery) the Confederate field pieces outperformed their Federal counterparts. The Southern battalions concentrated at vital points such as Hazel Grove, where "battery after battery wheeled into position to fire into Hooker's southern flank." Additionally, more than forty guns at Fairview pounded the Federal infantry and artillery around the Chancellorsville House (with other batteries firing from the southeast).[11] While the Confederate cannon inflicted heavy damage on the Federals, the Union batteries remained relatively ineffective in their response. As Douglas Southall Freeman stated, "For once [the Confederates] were fighting on equal terms against an adversary who on fields unnumbered had enjoyed indisputable superiority in weapons and ammunition."[12]

The Confederate cannoneers performed brilliantly at Chancellorsville; the new battalion system proved its worth.[13] After the battle, however, General Lee needed to make some changes that decreased the system's overall effectiveness – the artillery reserve was to be dissolved. Due to Lieutenant General Thomas "Stone-

---

[10] Longstreet's divisions were detached to guard the Virginia-Carolina coast. Vincent J. Esposito, ed. *West Point Atlas of American Wars,* Vol. 1, Map 84. (All references to Volume 1).

[11] Furgurson, *Chancellorsville 1863,* 227, 246 – 247.

[12] Freeman, *Lee's Lieutenants*, Vol. 2, 592.

[13] McWhincy and Jamieson, *Attack and Die,* 119. "The Confederates achieved one of their greatest successes with artillery at Chancellorsville." ; Furgurson, *Chancellorsville, 1863,* 227. "...Confederate guns quickly massed for their finest day of the entire war." ; Freeman, *Lee's Lieutenants,* Vol. II, 592. "At Hazel Grove, in short, the finest artillerists of the Army of Northern Virginia were having their greatest day."

wall" Jackson's death, Lee decided to reorganize his army by creating a third corps. This new command required that additional artillery battalions be assigned to it.[14] In order to supply these batteries Lee eliminated the artillery reserve, redistributing the units. Additionally, Lee's Chief of Artillery, General Pendleton (in whom few Confederate officers had much confidence), lost tactical control of the artillery; his duties now became essentially administrative.[15]

Each of the three corps in the Army of Northern Virginia had one artillery battalion assigned to each division, with two more forming the corps reserve. The battalions were placed under the command of the corps chief of artillery, who reported directly to the corps commander. The tactical control of the army's field pieces was now in the hands of the chief of artillery for each corps.[16]

One advantage of this reorganization for the artillerists was an even better promotion scale. This meant the ranks of the Confederate battalion commanders and reserve officers began to match their counterparts in the infantry. With this, many of the highly skilled officers could remain in the artillery and still gain their promotions.[17] Ironically, these changes were exactly what the Federal artillery commander, Brigadier General Henry Hunt, had long been lobbying for in the Federal army.

The poor performance of the Federal cannon at Chancellorsville gave Hunt the leverage he needed for making vital reforms. Before the battle, Major General Joseph Hooker made some ill-advised changes in the artillery command and organization. His

---

[14] E.P. Alexander, *Military Memoirs of a Confederate*, (New York, 1907), 370. Alexander stated that the reserve was "discontinued due to our poverty of guns, not to dissatisfaction with the system. And the fine service at Gettysburg by the Federal reserve of 110 guns, under Hunt, would seem to demonstrate the advantage of such an organization in every large army."

[15] Jeffrey Wert, "'Old Artillery': William Nelson Pendleton" *Civil War Times Illustrated*, June 1974, 17; Wise, *The Long Arm of Lee*, 570.

[16] Coddington, *The Gettysburg Campaign*, 13; Wise, *The Long Arm of Lee*, 570.

[17] Coddington, *The Gettysburg Campaign*, 13.

system proved to be ineffective and directly contributed to the mishandling of the Northern guns during the battle.[18]

Aside from the lack of centralized control, the Northern batteries had other problems. Not only did Hooker leave many of his guns in the rear when he crossed the Rappahannock, but also the batteries that did advance were poorly supplied and had difficulty replenishing their ammunition. Chancellorsville was one of the worst performances for the Federal artillery of the entire war. Most of the problems could not be attributed to the individual battery commanders, but were due to the faulty organization and handling of the cannon by Hooker and his staff.[19] Fortunately for the Federals, dramatic changes were made that greatly improved the effectiveness of the Northern guns at Gettysburg.

Following the battle, Hunt received permission to establish a brigade system (much like the Confederate battalion command). One of Hunt's subordinates, Colonel Charles Wainwright, initially conceived the idea of a brigade system months before. Hunt now had the opportunity to implement his plan.[20]

A post of chief of artillery was created for each of the seven infantry corps (also referred to as the artillery brigade commander). These officers were assigned an artillery brigade, most consisting of five batteries (see Appendix Two for details). Hunt

---

[18] Edward G. Longacre, *The Man Behind the Guns: A Biography of Henry Jackson Hunt, Chief of Artillery, Army of the Potomac* (New York, 1977), 150. General Hooker created a decentralized command system, giving the divisional and corps commanders more direction over their deployment. Hooker was an advocate of the decentralized use of artillery where the guns were spread out to bolster the infantry's line, and thus giving the local commander authority over the placement of the batteries. (see Sears, *Chancellorsville,* 68). Additionally, one of Hooker's biggest mistakes was in taking the tactical control of the guns away from Hunt.

[19] General Hunt said Hooker's meddling with the artillery "resulted in such mismanagement and confusion [that Hooker] consented to organize the artillery into brigades" after the battle. Robert Underwood Johnson and Clarence Clough Buel (eds.), *Battles and Leaders of the Civil War* (New York: The Century Company, 1887), Vol. III, 259; cited hereinafter as *Battles and Leaders.*

[20] Longacre, *The Man Behind the Guns*, 125.

made sure the brigades were attached to the corps and not to divisions or infantry brigades, thus keeping them from being subject to the whims of local infantry officers. This new system added flexibility and increased the potential for coordinating and concentrating guns on the field.[21] Additionally, Hunt arranged for the artillery commanders to have jurisdiction over the ammunition and supply trains (hoping to avoid the problems experienced at Chancellorsville).[22]

Thus, by June of 1863, both armies had organized their batteries into efficient commands. The arrangements each chose were remarkably similar. The main difference between the two, however, was that the Federals maintained a large artillery reserve, while the Confederate reserve was eliminated. Additionally, the Federals maintained an edge in the number of guns, uniformity of their batteries, and quality of their ammunition.[23]

[21] Coddington, *The Gettysburg Campaign*, 41 - 42; Naisawald, *Grape and Canister*, 329 - 330.

[22] Coddington, *The Gettysburg Campaign*, 42.

[23] Coddington, *The Gettysburg Campaign*, 250. To this point in the war the Confederate ammunition had been notoriously bad, but through the efforts of a very competent chief of ordinance and improved manufacturing, the ordinance hauled north by the Confederates came close to that which the Federals carried in their limbers.

Chapter Three

The First Contest for McPherson Ridge

On June 30, 1863, Brigadier General John Buford's cavalry division rode into Gettysburg. Buford's command included two mounted brigades and an artillery battery – Lieutenant John H. Calef's 2nd U.S. Artillery, Battery A. The veteran cavalryman quickly appreciated the value of holding the small town, mainly because the terrain south of the village looked like excellent ground from which to fight.

The next morning, July 1st, Buford's two cavalry brigades and Calef's battery deployed west of the village (additionally, men were posted to guard against the approach from the north). Planning to dispute every inch of ground, Buford advanced his dismounted troopers so they could fire on the Confederates and fall back when the pressure on their front intensified.

As the dismounted cavalrymen moved forward, Colonel William Gamble, the commander of Buford's First Brigade, told Lieutenant Calef to use his own discretion in selecting the best ground for placing his six guns.[1] Calef scanned the terrain and opted to deploy his three-inch ordnance rifles on McPherson Ridge, just north of the Chambersburg Pike. With all three sections together on the elevated ground, Calef could concentrate his

---

1 *O.R.*, Vol. 27, Pt. I, 1030; Edward G. Longacre, *The Cavalry at Gettysburg* (Cranbury, N.J., 1986), 180-184.

fire on approaching troops or enemy batteries.

Responding to the lieutenant's commands, the drivers directed their six-horse teams and limbers toward McPherson's farm. On reaching the crest, the crews moved off the road and prepared to unhitch the pieces. Just then a courier brought Calef word that General Buford wanted to see him. The young artillerist, a recent graduate from the United States Military Academy at West Point, rode off to find the general.[2]

On reporting to Buford, Calef received instructions as to how the general wanted the six field pieces deployed. One section was to be placed on each side of the road and the third moved further to the south, covering the open ground beyond McPherson's woods.[3]

These orders showed a marked difference in Buford's and Calef's tactical objectives. As the battery commander, Calef wanted his guns concentrated in one location where they could coordinate their fire on an enemy battery or advancing troops. General Buford, on the other hand, wanted the sections dispersed, boosting the firepower of his thinly stretched line. The general might also have been trying to fool the Confederates into thinking he had more than one battery.[4] While these dispositions were tactically sound for Buford's overall objective (attempting to delay the Confederates until the Federal infantry arrived), it meant that coordinated fire for Calef's sections would be more difficult.[5]

Calef turned and rode back to his battery. Arriving at the top of the knoll, he shouted the new instructions: Lieutenant Roder's section would remain in position north of the pike; Sergeant Newman needed to move the left section across the road (toward the barn); and Sergeant Pergel would lead the center section south across the farm and deploy beyond the woods. Scrambling to their assignments, the men hitched the pieces to the limbers and moved

---

[2] *O.R.*, Vol. 27, Pt. I, 1030; Naisawald, *Grape and Canister*, 339 - 340.

[3] *Ibid.*

[4] Richard S. Shue, *Morning at Willoughby Run, July 1, 1863*, (Gettysburg, Pa., 1995), 81.

[5] *Ibid.* This dispersion also forced the young battery commander to ride back and forth over the quarter mile distance in which his guns were spread. "Overall, the battery line would then span 600 yards."

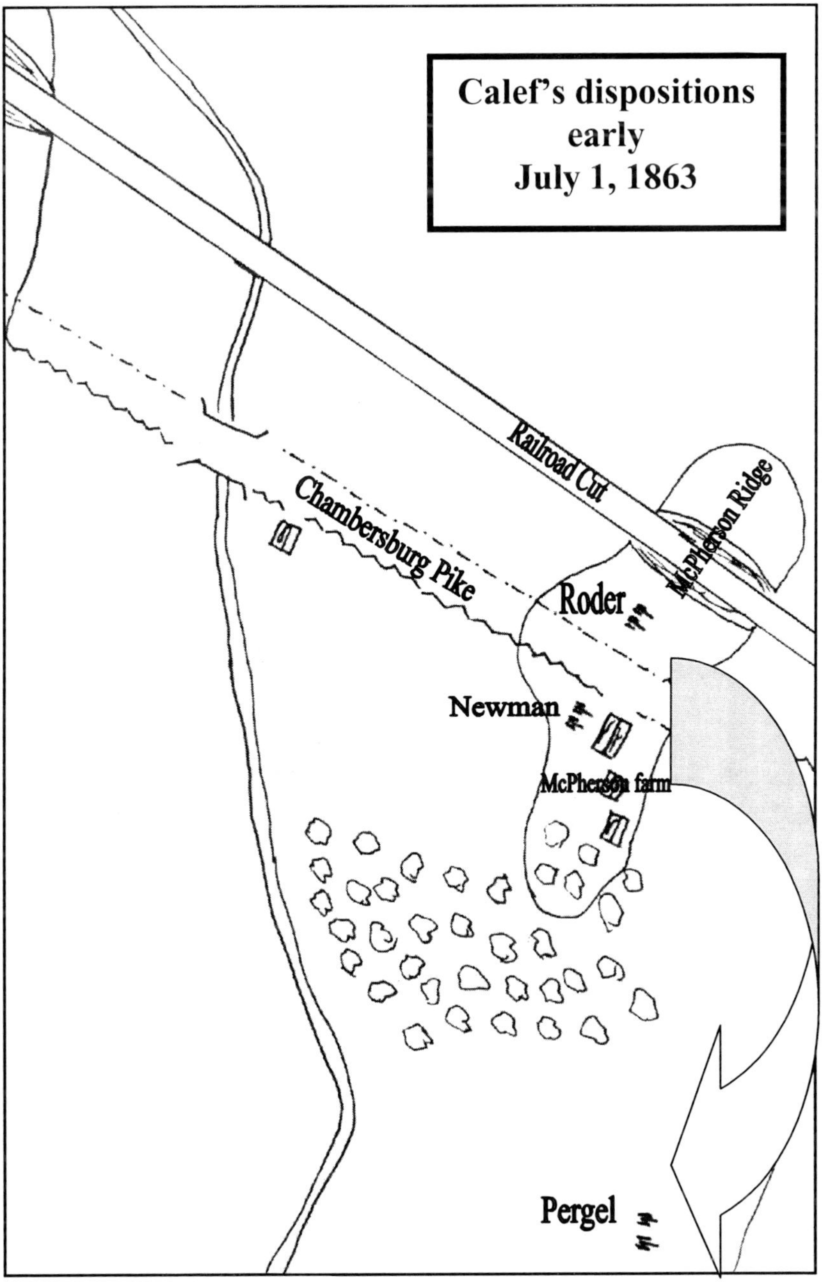
Calef's dispositions
early
July 1, 1863
Railroad Cut
McPherson Ridge
Chambersburg Pike
Roder
Newman
McPherson farm
Pergel

Colonel Charles Wainwright

Photograph from the U.S.M.H.I.

to their positions. Not long after Calef's instructions were given, Lieutenant Roder's section fired the initial Federal artillery shot of the battle.[6]

This first volley quickly drew the attention of the Confederate artillerymen on the ridge to the west, near the Belmont School House. Within minutes of Roder's first shot, Calef noticed a shell hurl overhead. The Southern guns replying to Roder's salvo were part of Major William. J. Pegram's artillery battalion. Pegram was an energetic officer, whom many considered to be one of the best artillerists in the service.[7]

Major Pegram became ill following the battle of Chancellorsville, only returning to duty on June 30 (the day before the first day's battle at Gettysburg). Seeing the artillerist back at his post was welcome news for the Confederate Third Corps. General Lee remarked to Lieutenant General A.P. Hill, "I have good news for you; Major Pegram is up." "Yes," Hill replied, "that is good news." The Third Corps commander later took time during the first day's battle to shake hands with this valued artilleryman.[8]

Pegram initially had difficulty finding a good location for his pieces – until Buford's skirmishers were driven back.[9] He then advanced and found elevated ground just west of Herr Ridge – giving his men a view of McPherson's farm and Calef's two sections. Once the guns were unlimbered, Pegram ordered at least two of his batteries to concentrate their fire on the four Federal guns near the farm. Outnumbered, Calef's two sections held their own. The batteries exchanged fire for over an hour, until advancing infantry threatened Calef's position.[10]

---

[6] *O.R.*, Vol. 27, Part I, 1031.

[7] Glenn Tucker, *High Tide at Gettysburg: The Campaign in Pennsylvania* (Bobbs-Merrill Co. Inc., 1958), 105. "one of the best in the service..."; Freeman, *Lee's Lieutenants*, Vol. 3, 197. Freeman stated, "Those who knew the younger artillerist ranked...Pegram...with Pelham." Pelham, the famous innovative young artillerist who used his flying artillery with much success, was killed three months earlier at Kelly's Ford.

[8] Freeman, *Lee's Lieutenants*, Vol. 3, 179 footnote.

[9] Wise, *The Long Arm of Lee*, 616.

[10] *O.R.*, Vol. 27, Part I, 1031 Lieutenant Calef stated in his official report, "Two of the enemy's batteries, one on each side of the road,"

Major General Henry Heth was in command of the Confederates advancing on Gettysburg from the west (Heth's Division of Hill's Corps). Expecting little opposition, General Heth planned to sweep a token force of militia from his front and then enter the village. Because he expected minimal resistance, Heth saw no need to wait for Pegram's field pieces to soften the enemy position. Thus, the Confederate batteries were left to fire as they pleased without any specific plan or objective.[11]

A spirited artillery duel between Calef's and Pegram's guns continued while Heth's infantry gradually drove the Union troopers back. Those Southern cannon not involved in the counterbattery action fired over the heads of the gray lines, hoping to assist with their attack. The elevated ground and the corresponding valley along Willoughby Run gave the artillerists the seldom achieved opportunity to support infantry on the offensive.[12]

While Heth's troops steadily gained ground, the lead elements of the Army of the Potomac, under the command of Major General John Reynolds, began arriving on the field southwest of the village. General Reynolds rode ahead to assess the situation and then went back to give instructions to his troops. Temporarily halting the head of the column, he ordered them to cut a diagonal course from their position south of the village to the northwest – bringing them to Buford's assistance more quickly.[13]

Part of this force was Captain James Hall's Second Maine

---

opened on his four guns; General Hunt wrote, "Calef's battery, engaging double the number of its own guns, was served with an efficiency worthy of its former reputation as 'Duncan's battery' in the Mexican War, and so enabled the cavalry to hold their long line for two hours." Henry Hunt, "The First Day at Gettysburg" from Battles and Leaders, Vol. III, 276.

[11] Wise, *The Long Arm of Lee*, 616.

[12] During the Civil War, artillery was usually relegated to mainly a defensive roll. McWhiney and Jamieson, *Attack and Die*, 116 – 117. "Tactical theory before the Civil War regarded artillery as equally an offensive and defensive arm, but Civil War experience proved otherwise."; Griffith, *Battle Tactics of the Civil War*, 171. "...primarily defensive functions, and it may be said that the success of Civil War artillery was usually much greater in defense than in the attack."

[13] *O.R.*, Vol. 27, Part I, 265; A.P. Smith, *History of the Seventy-Sixth Regiment New York Volunteers* (Cortland, New York, 1867), 236.

battery (six three-inch rifles). While more direct, this path required Hall's six guns to travel "through fields, gardens and yards" on their way to McPherson Ridge.[14] The cannoneers had some difficulties negotiating the numerous obstacles along the way; the fences, ditches and other obstructions slowed their progress.[15] Determination persevered, however, as the Maine men brought their guns through the tangle without mishap.

The approach of the First Corps infantry was a welcome sight to Buford's cavalrymen. The Confederates were gaining ground in several places, and the thinly spread troopers were tiring and running low on ammunition. With the enemy advancing dangerously close, General Buford ordered Calef to withdraw his guns. The lieutenant was instructed to remove one piece per section and then follow with the second. This maneuver was an attempt to reduce the vulnerability of the battery as it withdrew.[16] Rather than having both pieces of a section defenseless at the same time, one would stay and fire while the other hitched to the limber and started for the rear. Unfortunately, the last gun to leave would not have the benefit of cover fire.

On the south side of the pike Sergeant Newman ordered his first gun to head for the rear. Rushing the limber to the piece, the crew quickly had it hitched and thundered off toward the Seminary. Once the first gun was safely out of the line of fire, the remaining limber was ordered up. At this point disaster struck – four of the six horses in the team were cut down by an exploding shell. The crew was in a desperate situation; they needed to get word back to the caissons for another limber to come forward to take the piece off the field. This would take time, and the advancing Confederates might overtake the gun before the limber arrived.

Sergeant Newman and his artillerists met the challenge with an impressive display of determination and bravery. They scram-

---

[14] Smith, *History of the Seventy-Sixth*, 237.

[15] A member of the 147th New York remembered that the fence across the Fairfield Road especially "hindered the battery." William F. Fox, *New York at Gettysburg*, [Albany, 1900), Vol. III, 990. Cited hereinafter as *New York at Gettysburg.*

[16] *O.R.*, Part I, 1031.

bled to cut the harnesses that bound the dead horses to the limber. Once freed, they used the two surviving animals to extricate the gun.[17]

With Newman's and Roder's sections safe, Lieutenant Calef rode several hundred yards south to check on his remaining section. He arrived just in time as the cavalry support had already fallen back, and the Southern troops could be seen emerging from the woods. Knowing the limitations of two field pieces firing at scattered troops – especially when those pieces were three-inch rifles – Calef ordered the section to withdraw immediately.

While Buford's troopers and Calef's guns were retiring, the infantry from General Reynolds' First Corps began to deploy. The massed lines of Union troops were inviting targets for the Confederate cannoneers on the next ridge. With no enemy artillery to distract them, Pegram's batteries could concentrate on the Union infantry. Seeing the danger, General Reynolds instructed Captain James Hall to occupy the ridge just north of the Chambersburg Pike and engage the Southern artillery. Reynolds then ordered Brigadier General James Wadsworth, the commander of the First Division of the First Corps, to place infantry support on each side of Hall's battery. Turning to ride off, Reynolds said to Hall, "I desire you to damage the Artillery to the greatest possible extent, and keep their fire from our infantry until they are deployed, when I will retire you somewhat as you are too far advanced for the general line."[18]

Reynolds was using Hall's battery to draw some of the artillery fire off his infantry.[19] The general, a former artillerist from the Mexican War, knew exactly what he was doing by sending Hall to occupy the exposed ground. The battery would draw the attention of some of the enemy cannon and thus lessen the pressure on his infantry.

---

[17] *Ibid*; Shue, *Morning at Willoughby Run*, 101.

[18] Letters of James A. Hall to Colonel John B. Bachelder, February 27, 1867 and December 29, 1869. These letters are contained in *The Bachalder Papers*, edited by David and Audrey Ladd. (Dayton, Ohio: Morningside Press, 1994), 305-307 and 385-389. Citied hereinafter as *Bachalder Papers*.

[19] Naisawald, *Grape and Canister*, 342.

Captain Hall rode back to his battery and then led them toward McPherson Ridge. Approaching the elevation, Hall scanned the terrain and made a tactical decision that he hoped would minimize the dangers of occupying the crest. Rather than having his limbers exposed to counterbattery fire, he decided to move them further down the slope "to the foot of the crest."[20] Regulations called for the limber's lead horses to be six yards behind the tail of the gun.[21] Hall's decision meant that the number five man in each crew would have a longer distance to travel between the cannon and limber (he would also have to walk up and down the slope). Additionally, the increased distance would make it more difficult for the number six man to receive the gunner's instructions concerning ordnance selection and fuse lengths. Hall weighed this with the benefit of protecting his horses and ammunition chests, and decided the advantage of using the slope was worth the potential decrease in efficiency.

Quickly loading their pieces, Hall's gunners began dueling with the Southern cannon on the next ridge. The Second Maine claimed their first salvo of shot and shell caused the Confederates to move two of their pieces under cover. The opposing guns continued to fire at each other for about twenty-five minutes, when Hall's battery suddenly began receiving small arms fire from their front and right flank.[22]

Unknown to Captain Hall, there was a hidden railroad bed just north of his position. Advancing undetected down the cut, the skirmishers began picking off Hall's men and horses. Riding to his right to investigate, the captain discovered the threat. The right section was ordered to change front and face north to protect the battery.

While his crews changed positions, the gunners from the right and middle sections shouted new orders back to the limbers;

---

[20] Hall to Bachelder, December 29, 1869. *Bachelder papers*, 386.

[21] Jack Coggins, *Arms and Equipment of the Civil War*, (New York, 1983), 64 -65; Pfanz, *Gettysburg: The Second Day*, 491n.

[22] *O.R.*, Vol. 27, Part I, 359. "...our first six shots causing the enemy to change the position of two of his guns and place them under cover behind a barn."

rounds of canister were to be brought forward. The left section continued dueling with the Southern batteries while the other four guns tried to deal with the enemy infantry.[23]

Hall hoped to buy time for his battery, but infantry support was needed if he was to remain on the knoll. The initial blasts of canister forced the butternut and gray troops back to the cut for protection. Unfortunately, the relief was short-lived, and before long more of Hall's men and horses were casualties. Spreading their lines, the Confederates made themselves less vulnerable to the canister fire. One Federal infantryman who witnessed this action stated that artillery firing at "skirmishers is like shooting mosquitoes with a rifle."[24]

Hall realized his position was quickly becoming untenable. He concluded that the best way to extricate his guns would be to withdraw them by section. The first section to move would be the one closest to the railroad cut – the one in the most imminent danger. The lieutenant in command of the section, William Ulmer, was ordered to move back a short distance towards town. Rather than sending the section back to Seminary Ridge, Hall directed Ulmer to occupy an elevation approximately seventy-five yards to their rear. From here, Hall hoped Ulmer's section could aim down the cut with a good angle of fire at the enemy skirmishers.[25]

This was a sound tactical decision, but unknown to Hall the Confederates were already too far beyond his right flank for the plan to have worked. After Ulmer's teams left the knoll they received a surprising amount of small arms fire. Many of the Confederates had advanced further down the cut and leveled their muskets on the inviting targets. Once Ulmer's guns raced past, the butternut and gray troops followed the cannon toward town.[26]

On reaching the next crest, Lieutenant Ulmer's section halted and prepared for action. They immediately came under a savage storm of lead, "and a raking fire [began] sweeping over his guns

---

[23] Hall to Bachelder, December 29, 1869. *Bachelder Papers*, 386.
[24] *New York at Gettysburg*, Vol. 3, 991.
[25] Hall to Bachelder, December 29, 1869. *Bachelder Papers*, 386.
[26] *Ibid.*

so that he did not fire a shot." Every horse in one team was hit.[27] Rather than assisting the battery's other two sections, Ulmer's crews now had to work to save their own guns from capture. Cutting the dead horses loose, they managed to drag the piece off the field by hand.[28] Captain Hall then realized these cannon would be of no help in covering his guns.

With his initial plan no longer viable, the captain quickly thought of another strategy. Ordering his men to continue firing as rapidly as possible, he rode down the slope to speak with the limber crews. They were instructed to reverse front, facing the village and to be ready to move at a moment's notice. Rather than bringing the limbers into the heavier fire on the ridge, Hall planned to bring the guns to them. With the teams in motion, Captain Hall reined his horse and headed back up the hill to inform the men of his plan. They were directed to fire their last round and "under cover of the smoke [take the guns] … down the slope by hand to limber." All four pieces fired and were pulled down the slope to the waiting teams. Once hitched, they immediately headed toward town.[29]

Racing down the grassy strip between the Chambersburg Pike and the railroad cut, they came under what Hall described as a "hellish" fire. Hoping to capture the guns, the gray skirmishers fired at the animals pulling the limbers. Frantically the horses pressed on.[30]

Incredibly all four gun crews made it through the slight depression between the ridges and then headed up the gradual incline to the crest where Ulmer tried to deploy. Once on the ridge, Hall decided to cross the fence just beyond his lieutenant's former position (apparently there was a low point or a gap here). During this maneuver the cannon and limbers were especially vulnerable, only one being able to clear the rails at a time. This delay afforded the Confederates an opportunity to make their way further down the

[27] *Ibid*. In his official report Hall stated that only four horses were shot; *O.R.*, Vol. 27, Part I, 359.
[28] *O.R.*, Vol. 27, Part I, 359.
[29] Hall to Bachelder, December 29, 1869. *Bachelder Papers*, 386-387.
[30] *Ibid*; *O.R.*, Vol. 27, Part I, 359.

cut, where they took careful aim and shot down the horses of Hall's last gun. With no time to cut the animals loose and drag the piece off by hand, the captain reluctantly left it behind.[31]

As the 2nd Maine limped back toward the Seminary, Hall could count but four serviceable guns. Aside from the piece that was abandoned, another was struck by a solid shot from one of Pegram's guns as it retired; the cannon slowly moved toward the Seminary with a broken wheel. And Ulmer's piece, previously dragged to the road by hand, was hitched to a caisson limber and hauled to the rear.[32]

At this point the greatly discouraged captain met General Wadsworth on the road. Hall said he felt it was "cowardly" for his battery to have been isolated without proper infantry support. Apparently ignoring the brashness of the captain's language, Wadsworth informed Hall of Reynolds' death. He then instructed the captain to continue with his battery to the rear. Protesting, Hall asked for permission to retrieve his abandoned gun. Wadsworth reiterated that the battery was to move to the rear and to waste no time in doing it.[33] Before moving back, however, Hall dispatched a sergeant and five men to try to reclaim the piece. These men were all either captured or wounded.[34] The fate of the 2nd Maine's cannon now rested in the hands of Wadsworth's infantry, who were at that very moment moving to counter the advance of the Confederates in the railroad cut.

Fortunately for the Federals, the initial Southern assault was met by the First Corps infantry, who drove back and captured many of the Confederates (both in the railroad cut and near Willoughby Run). Following this effort a temporary lull settled over

---

[31] *O.R.*, Vol. 27, Part I, 359. James Hall's letter to the Adjutant General of Maine on July 14, 1863. "The boys fought like the devil, never better. You may judge when I tell you that many of our horses were not shot but bayoneted...I have seen hard fighting before and have been badly smashed but I never saw a battery taken from the field and its guns saved in so bad a state as the Old Second came off that day."

[32] Hall to Bachelder, December 29, 1869. *Bachelder Papers*, 387.

[33] Hall to Bachelder, February 27, 1867 and December 29, 1869, *Bachelder Papers*, 306-307 and 387.

[34] *O.R.*, Vol. 27, Part I, 359.

the field.

To this point in the battle, neither side's artillery had been utilized to their fullest potential. The Confederate batteries lacked specific tactical objectives – mainly because General Heth did not realize the quality of the opponents on his front.

The Federal batteries were also limited in their effectiveness. One problem for the Northern guns was the exposed positions which they occupied. This cost Calef and Hall men, horses, and the temporary loss of one field piece and two artillery limbers.[35]

Another problem was that both Calef's and Hall's batteries contained three-inch rifles, better suited for counterbattery fire than for trying to hold off infantry at close range. These batteries were not driven from the field by artillery fire, but by advancing infantry.[36]

Both sides regrouped, preparing for the next phase of battle, when the artillery would play a more prominent role in the action.

---

[35] One can safely assume that an artillery limber was lost with the gun that Hall left on the field (due to the loss of all of its horses). Also, if Ulmer's gun lost all six of his horses, then his limber must have been left on the field when the men hauled off the gun by hand. *O.R.*, Vol. 27, Part I, 359.

[36] This does not mean that 12-pound Napoleons could have held off the infantry attacks, but merely points out that rifled pieces were not as effective against infantry at close range as the larger bore Napoleons.

Chapter Four

The Confederate Batteries Make Their Presence Known

After the initial repulse on McPherson Ridge, General Henry Heth began making plans to renew his attack. The preparations included an impressive array of artillery firepower; nine batteries from Major W.J. Pegram's and Major D.G. McIntosh's battalions were in position to shell the Federals. Pegram's batteries rolled into line just south of the Cashtown Road, and Captain R.S. Rice's four twelve-pounder Napoleons and a section of Captain W.B. Hurt's long range rifled Whitworths (from McIntosh's battalion) joined them. Just across the road, Lieutenant Samuel Wallace's two sections of three-inch rifles readied for action. Six other guns moved several hundred yards to the south, near the Fairfield Road, where they came into battery on a hill with a commanding view. Once in place the Confederate cannon opened with a slow deliberate fire on the Federal positions.[1]

About this time three of the four remaining batteries in the Army of the Potomac's First Corps artillery battalion, commanded by Colonel Charles S. Wainwright, arrived. Wainwright's artillery brigade consisted of three batteries of three-inch rifles (two bat-

[1] Wise, *Long Arm of Lee*, 618 - 619.

teries with six rifles and the other with four) and two batteries of twelve-pounder Napoleons (six guns each).[2]

Colonel Wainwright left his artillery command on Seminary Ridge, and rode forward to reconnoiter the ground. During this inspection he received a message from Major General Abner Doubleday (Doubleday assumed command of the First Corps after General Reynolds was killed). Doubleday wanted Wainwright to post a battery on the ridge by McPherson farm, "if possible." Sending word back to the Seminary for one of his batteries to come forward, the colonel rode ahead to examine the ground. The prospect of placing a battery on the exposed crest (the same one that Calef and Hall had previously occupied) concerned Wainwright, and he quickly decided against it. Riding back toward town he met Captain Gilbert Reynolds and his 1st New York Light Artillery, Battery L, coming forward in response to his orders. Wainwright briefly explained the situation and instructed the unit to turn around and head back.[3]

Continuing toward the Seminary, Wainwright noticed another battery moving out of town – Captain Hall's 2nd Maine. The colonel inquired as to their movement. Hall explained that General Wadsworth had ordered him back to the ridge. Wainwright immediately countermanded the order, rejecting the idea of re-occupying the exposed crest. In his journal Wainwright wrote, "I took the responsibility of forbidding him to put his battery there until I knew there were troops to cover his right flank." Hall then received permission to retrieve his abandoned piece before heading for the rear.[4]

Captain Hall gladly set about his mission and called for a limber from the Seminary. The men of the 6th Wisconsin, part of the Iron Brigade of Wadsworth's division, had previously driven the Confederates from the cannon and prevented its capture. A

---

[2] Busey and Martin, *Regimental Strengths and Losses at Gettysburg*, 30; *O.R.*, Vol. 27, Part I, 230; Warren W. Hassler Jr., *Crisis at the Crossroads: The First Day at Gettysburg* (University of Alabama Press, 1970), 56.

[3] *O.R.*, Vol. 27, Part I, 355.

[4] Allan Nevins, (ed.) *A Diary of Battle: The Personal Journals of Colonel Charles Wainwright, 1861 - 1865* (New York, 1962), 234. (Cited hereinafter as *A Diary of Battle*).

captain in the 6th ordered a detail to pull the gun closer to the road so the artillerists could reclaim it. Hall and his men "thank[ed] us a thousand times" for saving the piece, noted a Wisconsin private, and in return they took off a wounded sergeant for them.[5]

After Colonel Wainwright refused to place Hall's battery on the knoll, General Wadsworth, who must have been less than pleased at having his orders countermanded, found Lieutenant Calef and directed him to occupy the position. Since the 2nd U.S., Battery A, was not under Wainwright's command, the colonel could not interfere with this order. Calef attempted to voice his opposition, but Wadsworth would have none of it. Reluctantly the regular army officer led two of his serviceable sections back to McPherson Ridge.[6]

Colonel Wainwright was in a difficult situation. He needed to cooperate with Wadsworth in strengthening the infantry's defenses, but at the same time he wanted to insure the relative safety and effectiveness of his batteries. The key to Wainwright's dilemma centered on the fact he and Wadsworth had different objectives for the batteries.[7]

Infantry commanders were primarily interested in using field pieces for supporting their troops. This role often minimized the battery's overall effectiveness, however. For example, the decision to place a battery in a line of battle restricted its potential for advantageous deployment in other locations – where they might enfilade the enemy lines, for example. Additionally, placing a battery with the infantry usually limited its opportunities for counterbattery fire. Because infantry commanders naturally tried to avoid deploying on ground exposed to enemy artillery (a necessity for counterbattery positioning), the battery's effectiveness in dueling with the opposing guns could be reduced. Also, battery commanders were often discouraged from initiating counterbattery engagements because the enemy's return fire could inflict casualties on the nearby troops.

---

[5] Lance Herdegen and William J.K. Beaudot, *In the Bloody Railroad Cut at Gettysburg* (Dayton, Ohio, 1990), 207.

[6] *O.R.*, Vol. 27, Part I, 1031; Shue, *Morning at Willoughby Run*, 167.

[7] *A Diary of Battle*, 233-234; *O.R.*, Part I, 360.

When Calef's guns reached McPherson Ridge, the lieutenant ordered them into battery north of the road. Their presence drew the attention of Pegram's and McIntosh's guns across the valley, and twelve pieces opened on the 2nd U.S., Battery A.[8] A spirited duel ensued for at least fifteen minutes. The Confederates then shifted some additional guns further north, attempting to enfilade Calef's position. These guns probably belonged to Captain T.A. Brander's Letcher Artillery (two ten-pounder Parrott rifles and two twelve-pounder Napoleons).[9]

Noticing that Calef was outgunned, Colonel Wainwright reluctantly ordered a battery forward, "still thinking it an ugly place" for artillery.[10] Captain Gilbert Reynolds' Battery L, 1st New York Light Artillery (six three-inch rifles) moved up to assist Calef.[11]

The New York unit drew fire even before it came into battery. These projectiles, however, came from a new direction – the Federal right. This additional threat made Calef and Reynold's position untenable; both units were forced to retire from the ridge. The batteries moved approximately 500 yards to the southeast, where their left was protected from Pegram's and McIntosh's guns by the intervening woods.[12] The Federal cannon then focused on

---

[8] *O.R.*, Vol. 27, Part I, 1031-1032.

[9] Warren W. Hassler, *Crisis at the Crossroads: The First Day at Gettysburg* (University of Alabama Press, 1970), 32 and 110; *O.R.*, Vol. 27, Part II, 678; Busey and Martin, *Regimental Strengths and Losses at Gettysburg*, 193.

[10] From Wainwright's journal. This entry was not included in Nevins, *Diary of Battle*. Copies obtained from the Gettysburg National Military Park Library. Special thanks to Scott Hartwig and his staff for making these available. (Cited hereinafter as *Wainwright Manuscript*).

[11] *O.R.*, Vol. 27, Part I, 230 and 355; *New York at Gettysburg*, 1256 and 1257; Busey and Martin, *Regimental Strengths and Losses*, 30. It should be noted that Warren Hassler's *Crisis at the Crossroads* (page 56), and Richard Shue's, *Morning at Willoughby Run* (page 179) both state Reynolds battery as having only four guns. Hassler footnotes "The Reynolds Battery tablet, on the Gettysburg Battlefield." The previous sources, however, tend to suggest that Battery L did carry six guns into action that morning.

[12] *O.R.*, Vol. 27, Part I, 362. Wainwright's journal entry leads one to believe the two batteries (Calef and Reynolds') first traveled due south

the enemy artillery to the north.

The enfilade fire Reynolds and Calef received came from Oak Hill, approximately 1400 yards north of McPherson farm. The Confederate artillery on this elevation would play a vital role in the first day's battle. By occupying this knoll they not only threatened the Federal First Corps, but also later harassed the Eleventh Corps as they deployed in the fields north of town.

The guns shelling McPherson Ridge were part of Colonel Thomas H. Carter's artillery battalion, attached to Major General Robert Rodes' division. Educated at the Virginia Military Institute, Carter learned his gunnery skills under Professor Thomas Jackson (later known as "Stonewall" Jackson).[13] His command performed well at Chancellorsville where they occupied Hazel Grove and Fairfield, inflicting heavy damage on the Federals.[14]

Carter's battalion accompanied the lead elements of Lieutenant General Richard Ewell's Second Army Corps, arriving from the north. Colonel Carter moved part of his battalion to the open crest of the hill while the lines of Southern infantry prepared to deploy. These batteries were Captain W.P. Carter's (two ten-pounder Parrott rifles and two twelve-pounder Napoleons) and Captain C.W. Fry's (two ten-pounder Parrott rifles and two three-inch rifles). Colonel Carter had two additional batteries not yet engaged.[15]

The Confederate artillery on Oak Hill not only caused Calef and Reynolds to change position, but also created havoc for the Federal infantry in the area. One brigade commander commented

---

and remained on the McPherson Ridge. Calef's report tends to lend some credence to this as he states, "I then took up a position nearly on line with my former one, but better covered by a corner of woods from the batteries in front..." (*O.R.*, Vol. 27, Part I, 1032). However, after looking at other reports and additional evidence, including Wainwright's official report (where he states Reynolds' battery took a position "at right angles to the ridge..." *O.R.*, Vol. 27, Part I, 356), and Scott Hartwig's article, "The Defense of McPherson Ridge" in *The Gettysburg Magazine*, Issue One, July 1989, this was not the case.

[13]Fairfax Downey, *Guns at Gettysburg*, (New York, 1958), 45; Sifakis, *Who Was Who in the Civil War.*

[14] Furgurson, *Chancellorsville, 1863,* 246.

[15] *O.R.*, Vol. 27, Part II, 602; Hassler, *Crisis at the Crossroads*, 87; Busey and Martin, *Regimental Strengths and Losses*, 169.

that this fire made his "position hazardous and difficult in the extreme."[16] Several Federal regiments either changed front to the north, facing this fire, or moved under cover. Either way Carter's guns caused the Union dispositions to be adjusted, making them occupy ground that was less advantageous for defending against an attack from their west.[17]

Colonel Wainwright continued adjusting his batteries – attempting to deal with Carter's fire and in preparation for an expected attack from Hill's troops. Previously shifting Reynolds' guns so they faced Oak Hill, the colonel then rode over to Captain James Cooper's 1st Pennsylvania, Battery B (two sections of three-inch rifles). Wainwright initially sent the unit to support Federal infantry in a field several hundred yards south of the Cashtown Road. Cooper then received new orders: he was to move and assist Reynolds in his counterbattery effort against the Confederates to the north.[18]

While Reynolds' and Cooper's guns occupied the attention of the enemy cannon, Wainwright placed two additional batteries, Captain Greenleaf Steven's 5th Maine (six twelve-pounder Napoleons) and Lieutenant James Stewart's 4th U.S., Battery B (six twelve-pounder Napoleons), on Seminary Ridge. Because the guns in these units were smoothbores, with an effective range of about 1,200 yards, they were kept out of the long distance counterbattery duel.[19]

---

[16] *O.R.*, Vol. 27, Part I, 329.

[17] Brigadier General Lysander Cutler's brigade was caught in the artillery crossfire and was effectively neutralized. See Hassler, *Crisis at the Crossroads*, 89.

[18] *Ibid*, 87; Naisawald, *Grape and Canister*, 346 - 347.

[19] Coddington, *The Gettysburg Campaign*, 251. The author states that these guns had a maximum range of 1,680 yards but an effective range of around 1,200 yards.

Chapter Five

Dilger's Battery

While Colonel Carter's two batteries poured an enfilade fire into the ranks of the Federals along the ridges west of town, his two remaining batteries, Captain R.C.M. Page's (four twelve-pounder Napoleons) and Captain W.J. Reese's (four three-inch rifles), were brought to bear on the Eleventh Corps north of town.[1]

Major General Oliver Howard's Eleventh Corps began arriving south of Gettysburg at approximately 12:30 p.m. The Third Division, under the command of Major General Carl Schurz, was the lead element of the corps and marched onto the fields north of the village. Planning to establish contact with the right of the First Corps, Schurz's units soon came under severe fire from the Confederate artillery on and at the base of Oak Hill – especially from Page's Napoleons near the McLean barn.[2] The open ground compounded the severity of this fire, offering little or no protection from the shelling.[3] Page's and Reese's guns prevented the Eleventh and First Corps from making contact; thus, a gap existed in the Union line.[4]

The 1st Ohio Light Artillery, Battery I, under the command

---

[1] Hassler, *Crisis at the Crossroads*, 63; Busey and Martin, *Regimental Strengths and Losses*, 169; *O.R.*, Vol. 27, Part II, 603.
[2] D. Scott Hartwig, "The 11th Army Corps on July 1, 1863" *Gettysburg Magazine*, Issue Number two, January 1, 1990, 36 and 37.
[3] John M. Vanderslice, *Gettysburg Then and Now* (original 1897; Morningside Press reprint edition, 1983), 88; Hassler, *Crisis at the Crossroads*, 76.
[4] Wise, *The Long Arm of Lee*, 619; Hassler, *Crisis at the Crossroads*, 76.

of Captain Hubert Dilger, followed Schurz's division onto the field. These cannoneers previously saw action at Cross Keys, Second Bull Run and Chancellorsville, where the battery established itself as a first rate command. In fact, many considered Captain Dilger to be the best cannoneer in the Army of the Potomac.[5]

Dilger, a German native, attended the Karlsrule Military Academy and later served in the Grand Duke of Baden's horse artillery. Drawn to the war in North America, the artillerist asked for, and was granted, a leave of absence, enabling him to travel to the United States and join the Union army. At the Battle of Second Bull Run he solidly established a reputation as a brilliant commander. Then the following spring, at Chancellorsville, his bravery and innovative tactical handling of guns earned him the Medal of Honor.[6]

General Howard described Dilger as "one of those handsome, hearty, active young men that everybody liked to have near." Discarding the army issue trousers, Dilger adopted ones made of leather (because of their durability). This led to him being known throughout the army as "Leatherbreeches." A tough disciplinarian, the former horse artillerist demanded a lot from his men. They constantly improved their skills through drill and target practice. This training, coupled with Dilger's abilities, made the battery an efficient unit that performed well on the battlefield.[7]

Brigadier General Alexander Schimmelefennig, temporarily commanding the Third Division, ordered Dilger to place his guns somewhere between the Carlisle and Mummasburg Roads. The latitude that Schimmelefennig gave Dilger showed the confidence

---

[5] Wise, *The Long Arm of Lee*, 475. "regarded as one of the best artillery officers in the army.."; Furgurson, *Chancellorsville, 1863* 161. "...some considered him the best cannoneer in the Union Army"; Henry Kurtz, "Dilger's Battery," *Civil War Times Illustrated*, Vol. 1, Number 7, 39.

[6] Kenneth M. Kepf, "Dilger's Battery at Gettysburg," *The Gettysburg Magazine* January 1, 1991, Issue Number Four), 49 - 50; Carl Keyser, "Leatherbreeches at Chancellorsville" *Civil War Times Illustrated*, August, 1975. Captain Dilger did not receive the Medal of Honor until the war was over.

[7] Kurtz, "Dilger's Battery"; Kepf, "Dilger's Battery at Gettysburg"; Keyser, "Leatherbreeches at Chancellorsville".

Captain Hubert Dilger

1st Ohio Light Artillery, Battery I

Photograph from the U.S.M.H.I.

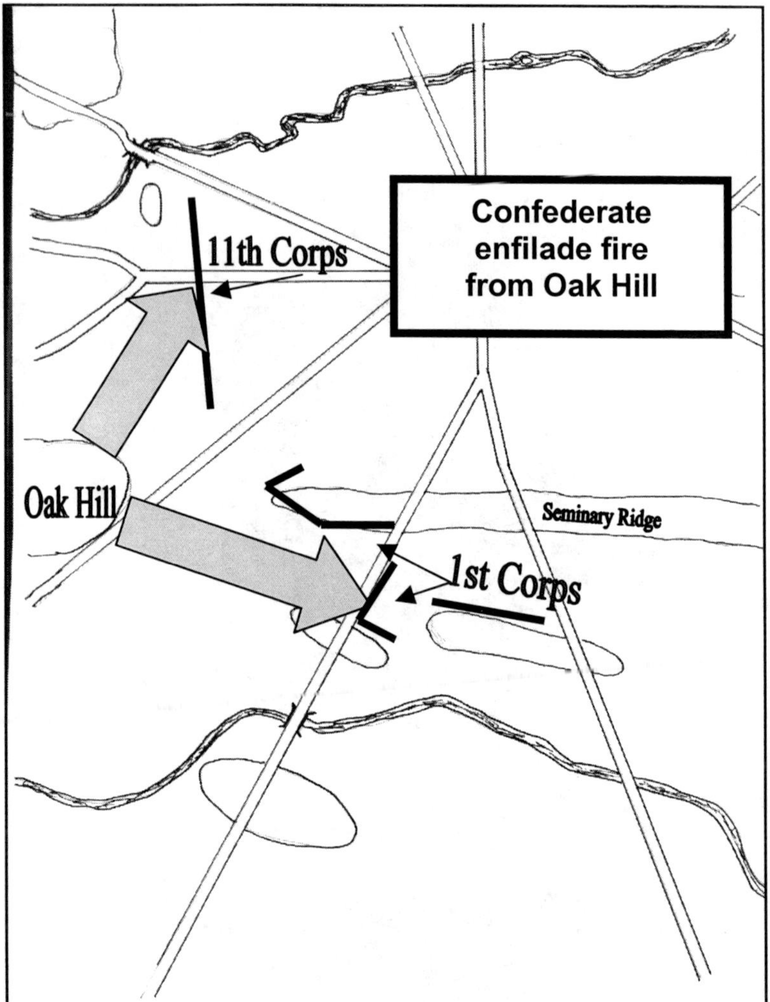
Confederate enfilade fire from Oak Hill
11th Corps
Oak Hill
Seminary Ridge
1st Corps

he had in the artillerist. After briefly scanning the terrain, Dilger sent one section to occupy an elevated position to his left, while he continued to reconnoiter. This section quickly drew counterbattery fire from Oak Hill, however, forcing him to rush his other sections forward for support.[8]

Dilger's six twelve-pounder Napoleons were soon exchanged with eight Confederate pieces 1,400 yards away. This distance was beyond the effective range for smoothbores, but Dilger's men rose to the challenge. With the battery's first shot sailing high, Captain Dilger then stepped forward and personally aimed the next round. This salvo dismounted a Confederate gun and killed some horses.[9] Continuing to pound Page's battery, their fire caused heavy damage and forced the Confederates to retire from the McLean barn. Page's guns came into battery again further up the eastern slope of Oak Hill.[10]

About this time Confederate General Robert Rodes began advancing his infantry. Colonel Edward O'Neal's Brigade moved south toward the salient occupied by Brigadier General John C. Robinson's division (Robinson held the northern end of the First Corps line). O'Neal's left received fire from four companies of skirmishers – part of the 45th New York. This fire did little to check the Confederate advance, so Captain Irsch, the commander of the skirmishers, sent word for Captain Dilger to assist if possible. Instructing the men of 45th New York to lie down, the Ohio

---

[8] Naisawald, *Grape and Canister*, 356; *O.R.*, Vol. 27, Part I, 754.

[9] Kepf, "Dilger's Battery at Gettysburg," 56; *O.R.*, Vol. 27, Part I, 754. "A four gun battery of the enemy immediately opened fire at about 1,400 yards on this section, and compelled me very soon to bring my whole battery into action. During this heavy artillery duel, the enemy had been re-enforced to eight pieces, of which two advanced [to within] 800 or 1,000 yards..."

[10] *O.R.*, Vol. 27, Part II, 603. "The artillery of the enemy [Dilger's battery]...delivered a very destructive oblique fire on Page's battery. His loss here was heavy - 2 men killed, 2 mortally wounded, 26 more or less badly wounded, and 17 horses killed and disabled."; Hartwig, "The 11th Army Corps on July 1, 1863" He cites Alfred Lee, "Reminiscences of the Gettysburg Battle," *Lippincott Magazine*, Vol. 6, (1883), 55; *O.R.*, Vol. 27, Part I, 754. Dilger claimed five carriages were put out of action.

battery fired over their prone lines into the advancing Confederates. A Federal officer noted that Dilger's battery fired "so successfully that the massed enemy began to halt and waver."[11]

Dilger's efforts drew the attention of General Rodes, who ordered additional pieces brought to bear on the Federal cannon. Captain W.P. Carter's battery (with two more rifled guns and a pair of Napoleons) assisted in this effort. "Here these batteries [the Confederate guns on the eastern slope of Oak Hill] rendered excellent service, driving back both infantry and artillery."[12]

With his unit outgunned, the German cannoneer sent word back to his chief of artillery, Major Thomas W. Osborn, to send up some rifled pieces. Osborn responded by bringing Lieutenant William Wheeler's 13th New York Light Artillery battery (four three-inch rifles) forward. He placed the two sections under Dilger's command.[13]

The 13th New York battery suffered greatly during the Chancellorsville campaign. They reluctantly abandoned two guns during the battle, and their commander, Captain Julius Dieckmann, resigned soon afterwards. Lieutenant Wheeler received command of the battery just prior to the Battle of Gettysburg.[14] Thus, Osborn's decision to place Wheeler's two sections under Dilger's watchful eye seemed very wise.

Only recently receiving the appointment as Eleventh Corps Chief of Artillery, Major Osborn seemed little impressed with his new command. "The artillery of this corps," he wrote, "has been

---

[11] *New York at Gettysburg*, 378 - 379.

[12] *O.R.*, Vol. 27, Part II, 603. Lieutenant Colonel Carter might have been exaggerating a bit when he implied the Federal artillery was driven off. Dilger in fact moved his guns into a better position to fire at these batteries.

[13] *Ibid.*, Part I, 754; Naisawald, *Grape and Canister*, 356; Herb Crumb, ed. *The Eleventh Corps Artillery at Gettysburg: The Papers of Major Thomas Ward Osborn, Chief of Artillery*. (Hamilton, New York, 1991), 2 and 85; Herb S. Crumb and Katherine Dhalle, eds., *No Middle Ground: Thomas Ward Osborn's Letters from the Field (1862 - 1864)*, 143 - 144. Citied hereinafter as *No Middle Ground*. Osborn was little impressed with the Eleventh Corps artillery, nor its commanders. Apparently Dilger was the only commander with whom he had any real confidence.

[14] *New York at Gettysburg*, 1320.

known throughout the army as the worst in the army."[15] The upstate New York lawyer decided to make the most of his new assignment, however, proving himself to be an insightful artillerist during the three days of battle in Pennsylvania.

Wheeler's guns unhitched just to the right of Dilger's.[16] The four rifles soon found their range, sending shot and shell toward the field pieces on Oak Hill. At this point Dilger boldly executed a move that demonstrated why so many of his peers respected his ability. Desiring to move his Napoleons and Wheeler's rifles closer to the enemy, the captain wanted to do so in a manner that minimized the risk for the crews, cannon and limbers. The problem was, the movement would create a lull in their firing that would allow the Confederate gunners time to load and aim without the distractions created by counterbattery fire. Additionally, when Dilger's pieces stopped firing, they would be more visible to the enemy (the guns would no longer produce smoke to screen their position). To overcome this danger, Dilger decided to advance only a few guns at a time. The remaining pieces would provide cover fire for those in motion – occupying the Southern cannoneers and providing a smoke screen to help conceal their initial movements.[17]

Dilger directed two crews to follow him forward under the cover provided by the remaining pieces. The captain selected a new position approximately 600 yards closer, where their fire would be more effective. Once established, he returned for the other sections. The next four guns moved to the new position under the protection provided by his advance pieces and Wheeler's guns. Dilger placed his six cannon in battery (600 yards closer to the enemy), and then sent for Wheeler's command.

With Dilger's cannon occupying the Confederates, Wheeler's guns advanced without mishap. Incredibly, all ten guns (both batteries) made the move without a loss. Making the advance even

---

[15] *No Middle Ground*, 143; *The Papers of Major Thomas Ward Osborn*, 2 & 85.

[16] *O.R.*, Vol. 27, Part I, 752.

[17] Murray, *Hurrah for the Ould Flag!*, 90. Andrew Cowan describes the lack of visibility created by the black powder charges.

more difficult was a five-foot ditch that had to be crossed; the first crews scrambled to fill the obstruction with fence rails and other material under heavy fire.[18] Without utilizing his innovative leap-frog tactics, it would have been difficult to move two batteries forward without losing a single piece.

Another tactic employed by Dilger worth noting was his plan for re-supplying the guns. Originally placing his caissons in the rear, out of the line of fire, he ordered three limbers to constantly shuttle ammunition to the front. This meant only one or possibly two of his caissons would be directly exposed to enemy projectiles at any point in time. Because Dilger's pieces fired in a slow and deliberate manner, they maintained a sufficient supply of ordnance.[19]

From his advanced position Dilger's guns engaged the enemy cannon with greater effect. Unfortunately, the German artillerist had little time to take advantage of the new location because his command came under fire from a different part of the field. Posted on a hill to Dilger's right, Confederate cannon sent shot and shell down his line of field pieces. This enfilade fire took a toll on the crews, horses, and guns (with one Napoleon being disabled). Compounding Dilger's problems was the presence of Confederate infantry moving beyond his flank. Lieutenant Wheeler's right section turned to face the new threat, pouring canister into their ranks.[20]

Realizing the futility of trying to remain in this position, Dilger gave the order to retire. Once again he utilized the tactic of leapfrogging his guns – using part of his battery to cover the movement of the other.[21] As the pieces from both Dilger's and Wheeler's units made their way back toward town, a solid shot bounded into the axle of one of Wheeler's gun carriages. In an attempt to salvage the dismounted piece, the lieutenant ordered the 800-pound tube tied under the limber and towed to safety. The

---

[18] *O.R.*, Vol. 27, Part I, 752, & 754.

[19] *Ibid.*, 754.

[20] *Ibid.*, 752-753, & 754.

[21] *Ibid.*, 754. Dilger stated that his "retreat was executed in the same manner as the advance..."; Naisawald, *Grape and Canister*, 359.

rope soon broke, however, and the gun was abandoned.[22]

Once he reached the edge of the village, Dilger deployed a section from each battery to fire at the approaching gray lines; the other pieces were sent ahead to make their way through the congested streets. The four remaining cannon pounded away at the oncoming Confederates, switching to canister when the troops closed to within a few hundred yards. Dilger soon ordered Lieutenant Wheeler's two remaining pieces to the rear, because these rifles at short range "commenced to become useless."[23] Remaining a few minutes longer than Wheeler, Dilger retired down Baltimore Street, most likely with prolong ropes tied so he could continue firing canister as they retreated.[24]

On the afternoon of July 1st, the German trained artillerist demonstrated a high degree of skill and tactical ability. Utilizing his guns to their fullest potential, he did so without foolishly jeopardizing them. The Confederates also demonstrated a great deal of skill handling their pieces, managing to enfilade both the First and Eleventh Corps positions with their batteries.

---

[22] *Ibid.*, 753.

[23] *Ibid.*, 754.

[24] *New York at Gettysburg,* Vol. III, 1320. Wheeler's battery – "Then with prolongs hitched so as to fire while retiring, the battery fell back to Cemetery Hill." It is difficult to say for certain whether or not Dilger used prolong ropes to drag his pieces behind his limbers, or if he just hitched them up and galloped down the street. Dilger's official report lacks specifics as to this move, but two things suggest that he might well have employed this technique. First, Sergeant Frederick Smith of Weidrich's battery, positioned on Cemetery Hill with a good view of town, later recalled Dilger's "Ohio Battery was coming up Baltimore street in column of pieces, with prolong rope fixed for firing in retreat." (*Ibid*, 1246). This account was read nearly twenty-six years later so its reliability is not beyond scrutiny. Second, this maneuver was sound procedure for this type of retreat and several other Union batteries did use this technique at Gettysburg. This suggests that the German artillerist used his prolong ropes in retiring his pieces.

Chapter Six

The Struggle North of Town

While part of General Rodes' division engaged the Eleventh Corps on the plains east of Oak Hill, Major General Jubal Early's division moved to threaten the Federal right (these were the troops Dilger saw advancing beyond his flank). "Old Jube" previously scouted south along the Harrisburg Road, ahead of his troops. Seeing the Union infantry and artillery on the plains north of town, he called his batteries to the front.[1]

Lieutenant Colonel Hilary P. Jones' battalion responded to Early's request. Jones placed twelve guns (Captain A.W. Garber's four twelve-pounder Napoleons, Captain C.A. Green's two ten-pounder Parrott and two three-inch rifles, and Captain W.A. Tanner's four three-inch rifles) on a slight knoll just east of the Harrisburg Road. The guns were about one-half mile north of the bridge over Rock Creek.[2] These cannon "opened a brisk fire upon the enemy's artillery and infantry, taking them in the flank as they were being massed upon Rodes' left and General Early's right."[3] From this strategic elevation, Jones' "guns were to play a significant role in the impending clash with the two Eleventh Corps divisions."[4]

---

[1] *O.R.*, Vol. 27, Part II, 495; Early to Bachelder, March 23, 1876. *Bachelder Papers*.

[2] *Bachelder Map*, "The First Day's Battle," 1876; Busey and Martin, *Regimental Strengths and Losses*, 162; Hassler, *Crisis at the Crossroads*, 77.

[3] *O.R.*, Vol. 27, Part II, 496.

[4] Hassler, *Crisis at the Crossroads*, 77

Brigadier Generals Alexander Schimmelfennig and Francis Barlow were in command of the two Federal divisions. Barlow marched out of the village with orders to anchor the Union right. Temporarily deploying near the county almshouse, he soon decided to advance toward an inviting knoll several hundred yards to his front.[5] Accompanying the division was the 4th U.S. Artillery, Battery G (six twelve-pound Napoleons). Barlow directed the nineteen-year-old commander of the battery, Lieutenant Bayard Wilkeson, to place his six cannon on the elevated ground that dominated the surrounding terrain.[6]

As young Wilkeson's guns came into battery, four companies of the 17th Connecticut were ordered to form a skirmish line in his front (the elevated ground allowed the cannon to safely fire over the troops). The Connecticut soldiers advanced down the slope and beyond Rock Creek, toward the Brenner house. Seeing their approach, Jones' gunners opened on them with "shot, shell, grape, and canister."[7] At least some of the cannon then focused on the Brenner house, causing the structure to catch fire and preventing the Federals from using it as cover.

Before long Jones' units focused their attention on Wilkeson's battery, clearly visible on the exposed knoll. Four of Battery G's guns were in action on the hill – one of the sections was detached to a position on the left.[8] The ensuing counterbattery

---

[5] Hassler, *Crisis at the Crossroads*, 69 -70.

[6] Naisawald, *Grape and Canister*, 357; Hassler, *Crisis at the Crossroads*, 69.

[7] *O.R.*, Vol. 27, Part I, 717. Although the contemporary used the term "grape," this was inaccurate, "for the use of grape in field guns had been discontinued in the United States service for a number of years." Naisawald, *Grape and Canister*, 541.

[8] *Bachelder Map*, "The First Day's Battle," 1876. After inspecting these maps Eugene Bancroft wrote Bachelder saying they were "wonderfully correct." Bancroft to Bachelder, May 28, 1878. *Bachelder Papers*; *O.R.*, Vol. 27, Part I, 757. Lieutenant Merkle, who commanded the detached sections, wrote that he moved "about one mile or three-quarters northwest of the poor-house." This placed him on Wilkeson's left as Blocher's Knoll, or as it was later known, "Barlow's Knoll," was almost due north of the county almshouse.

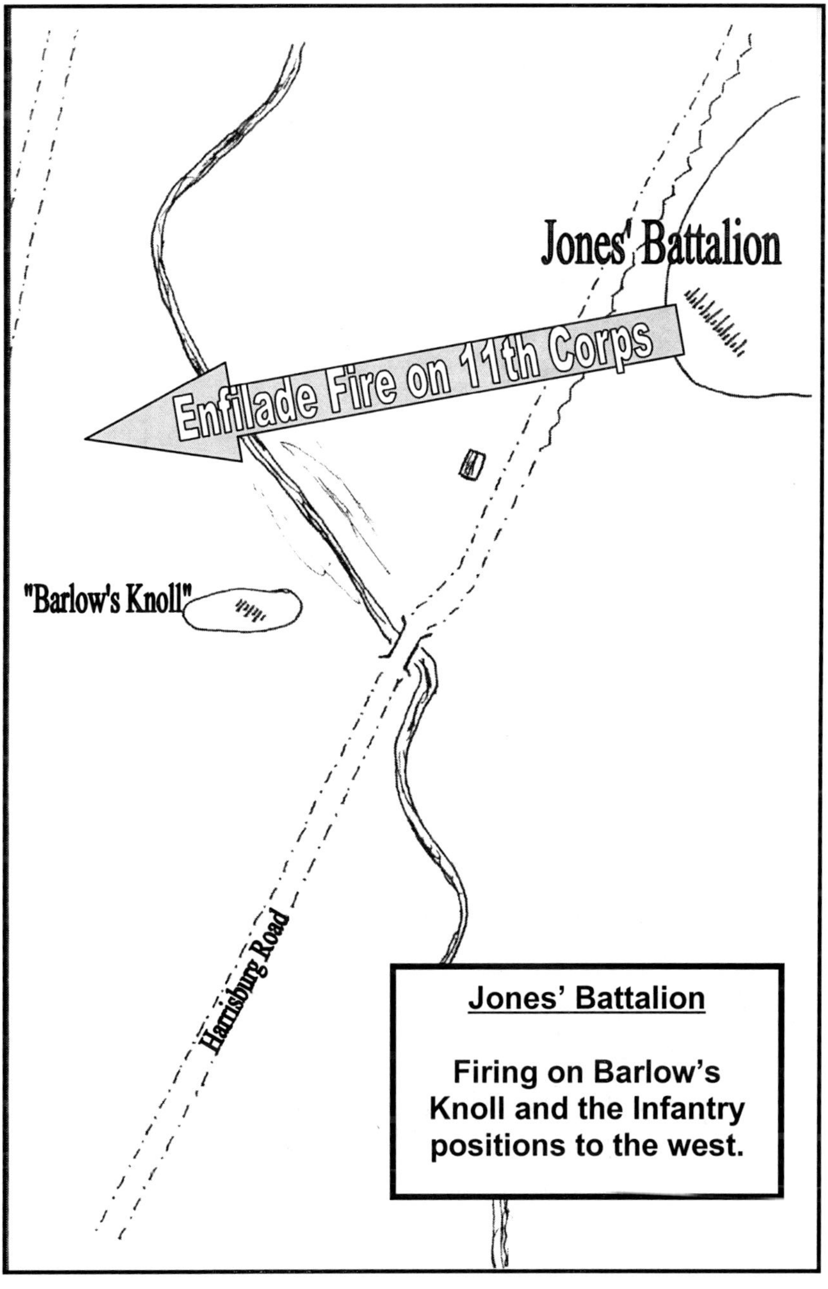
Jones' Battalion
Enfilade Fire on 11th Corps
"Barlow's Knoll"
Harrisburg Road
Jones' Battalion
Firing on Barlow's Knoll and the Infantry positions to the west.

Lt. Bayard Wilkeson in command of his battery before he was mortally wounded

exchange was short (only about one-half hour), but intense.[9] One of Battery G's solid shots struck a Napoleon in Captain Garber's battery; the projectile hit the face of the muzzle and bent the barrel. In the action Jones lost one man killed, one wounded, and two horses disabled.[10]

Incredibly, "three [of Jones'] guns were temporarily disabled by having shots wedged in the bores."[11] At least one of the cannon, a ten-pounder Parrott, was put out of service when a gunner mistakenly used ammunition intended for the three-inch rifles. The Parrott's bore was only 2.9 inches, and the three-inch round lodged in the barrel after being rammed part way down.[12] This problem was a symptom of the lack of uniformity in many Confederate batteries. The variety of guns necessitated separate ammunition chests for each type of cannon; in the heat of battle it was easy to use the wrong ordnance.

On the Federal side the losses were also heavy. Lieutenant Wilkeson suffered a severe wound from an exploding shell – he died later that day. Other casualties included one killed, four wounded, twelve horses killed and one gun disabled.[13] Major Osborn commented that Battery G was subjected to a "concentration of fire of which no battery could withstand."[14] Lieutenant Eugene Bancroft took command after Wilkeson went down. Bancroft soon ordered the two sections off the knoll, moving "a short distance" to escape the severity of fire.[15]

About this time Brigadier General John B. Gordon's Brigade, of Early's Division, crossed Rock Creek and drove the Federal

---

[9] *O.R.*, Vol. 27, Part I, 756.

[10] *Ibid.,* 495-497. The gun is assumed to have been in Captain Garber's battery because his was the only battery of the three on the hill that contained Napoleons.

[11] *Ibid.*, 495.

[12] *Ibid.*; James Hazlett, Edwin Olmstead, and M. Hume Parks, *Field Artillery Weapons of the Civil War*, by (University of Delaware Press, 1988).

[13] *O.R.*, Vol. 27, Part I, 756; Hartsig, "The Eleventh Army Corps on July 1, 1863," 43.

[14] *O.R.*, Vol. 27, Part I, 748.

[15] *Ibid.*, 756.

infantry back toward the hill. Brigadier General Harry T. Hays' Louisiana Tigers, along with Colonel Isaac E. Avery's regiments, pushed beyond the right of Barlow's division, threatening the Eleventh Corps rear. Many of the thinly stretched Northern units began to head for town.

Seeing the retreat, Bancroft ordered Battery G to the rear. He instructed the crews to retire with prolongs fixed, firing loads of double canister at the oncoming Confederates. Once the cannon reached the village streets they were quickly limbered and headed for Cemetery Hill.[16]

Several other Federal batteries provided cover fire during the retreat. As previously mentioned, Dilger's and Wheeler's guns paused to send blasts of shell and canister into the advancing gray ranks. Captain Lewis Heckman, in command of Battery K, 1st Ohio Light Artillery, received orders to assist the Eleventh Corps as they fell back. "He was then put into position," wrote Osborn, "with a view of holding the enemy in check until the corps had time to retire through the town...."[17] Heckman's four crews expended 113 rounds of ammunition in thirty minutes, most of this being deadly blasts of canister. General Hays, whose command was the target of most of these discharges, said, "the fire to which my command was subjected from the enemy's batteries...was unusually galling."[18] Although Heckman's rounds were affective, they could not hold back the assault for long. Osborn stated, "... though [Heckman] worked his battery to the best of his ability, the enemy crowded upon it, and was within his battery before he attempted to retire." When the captain gave the order to retreat, it was too late; the Confederates overran his position, capturing two of the pieces.[19]

Lewis Heckman's efforts are often overlooked and/or unappreciated. While losing guns was a disgraceful thing for an artillerist, he risked his pieces in an effort to buy time for his comrades who were trying to make their way through the congested streets.

---

[16] *O.R.*, Vol. 27, Part I, 757; *Bachelder Papers*, 1974.

[17] *O.R.*, Vol. 27, Part I, 748.

[18] *Ibid*, Part I, 755 and Part II, 479.

[19] *Ibid*, Part I, 748 and 755.

By remaining in position he delayed the Southern advance slightly, thus providing precious time for the retreat. Judging an artillerist's performance based solely on losing guns is sometimes unfair. As General Hunt noted, there are times when "guns can be lost with honor."[20]

Once Heckman's remaining pieces left their position near the college, they were still not out of danger. The Confederates followed them down the street, firing as they came. Heckman's guns then joined part of Wheeler's New York battery and the two units made for the town square. Major Osborn met them along the way and accompanied them as they tried to escape. Suddenly one of the limber wheels broke (probably a limber in Heckman's command). Osborn ordered the field pieces to unlimber and fire canister at the enemy while the remaining crewmembers repaired the carriage. They took the spare wheel off a caisson and within "two or three minutes" replaced the damaged wheel. This done, they resumed their retreat. Major Osborn was impressed with their coolness and efficiency under fire.[21]

When the Eleventh Corps units began to withdraw, Lieutenant Colonel Jones' battalion did not remain idle. Directing their fire "on the disordered masses of the enemy that were rapidly retreating before our troops," the Southern artillerists hoped to turn the movement into a rout. Jones received word from General Early to send some guns forward to lob shells at the retiring forces. Captain Carrington's battery of four Napoleons dashed across the Rock Creek bridge in pursuit. Before they could deploy, however, their infantry had driven the Federals through the town.[22]

---

[20] *New York at Gettysburg*, 1254. In his official report Major Osborn stated, "I think no censure can be attached to this battery for the loss of the gun[s]."; *O.R.*, Vol. 27, Part I, 748.

[21] *Osborn Papers,* 13. The assumption that it was one of Heckman's guns is based solely on the two battery commanders' reports. Wheeler's report was fairly detailed and most likely would have included this event if it was one of his pieces. Heckman's report was very short, and included little detail. Since Osborn states that it was one of these two units, and neither commander's report mentions the event, it was most likely one of Heckman's limbers.

[22] *O.R.*, Vol. 27, Part II, 495.

Jones' battalion played a major role in the Confederate offensive north of Gettysburg. From their elevated position they were able to enfilade the Federal lines, thus weakening the Eleventh Corps defenses. A prominent member of Lieutenant General Ewell's staff said that he never saw guns "better served than Jones' were on this occasion."[23]

[23] Coddington, *Gettysburg Campaign*, 291.

Chapter Seven

The Defensive Stand on Seminary Ridge

Elements of the Confederate Third Corps had initiated the battle west of town early on July 1. Later that afternoon, General A.P. Hill's Third Corps prepared to make their final push against the Federals defending the ridges west of town. This effort roughly coincided with Early's assault north of the village.

The Confederates massed an impressive number of artillery pieces from Pegram's and Major David G. McIntosh's battalions on Herr Ridge to prepare for the attack. These cannon were deployed in a line perpendicular to the road. Pegram probably had eight Napoleons and five rifled pieces in action (two howitzers were held back and a rifled piece in Zimmerman's battery was disabled as it came into position). McIntosh had six Napoleons, six three-inch rifles, and two long range Whitworth rifles. Further to the north, Captain T.A. Brander's battery (in Pegram's command) had two Napoleons and two ten-pounder Parrott rifles posted on some elevated ground east of Willoughby Run. Captain M. Johnson's four three-inch rifles and a section of Hurt's three-inch rifles moved southeast to a position on the Hagerstown Road but were not engaged.[1]

---

[1] *O.R.*, Vol. 27, Part II, 674-675 and 677-678. There are several conflicting accounts as to the exact dispositions. The *Bachelder Map* lists the batteries from south to north as follows: (South of the Chambersburg Pike) Lieutenant Samuel Wallace's two three-inch rifles and two

With Pegram's and McIntosh's guns in position on Herr Ridge, and Carter's battalion firing from Oak Hill, the Federals defending McPherson Ridge were caught in a crossfire. This prevented the Union commanders from making the types of dispositions they wanted for defending against Hill's forthcoming assault. The Northern lines were continually adjusted to minimize the enfilade fire from the Confederate artillery.[2]

Colonel Roy Stone's brigade of three Pennsylvania regiments was especially hard hit. Stone shifted the 149th Pennsylvania to face north (toward Oak Hill), in an attempt to reduce the effects of Carter's fire (with McPherson's Ridge partially shielding the lines from the projectiles from west). He then placed the remaining regiments, the 143rd and 150th Pennsylvania, on McPherson Ridge facing west, with the McPherson barn and house protecting their right flank.[3]

Despite these efforts, the 149th continued to take casualties. One round took the head off a soldier and a moment later another shell injured seven more. In an attempt to lessen the severity of the fire, the commander of the regiment, Lieutenant Colonel Walton Dwight, ordered the color guard to advance fifty yards. This decoy worked; the Confederate guns on Herr Ridge changed their angle of fire to where they believed the regiment had relocated.[4]

Napoleons, Captain R.S. Rice's four Napoleons, Captain E.A. Marye's two three-inch rifles and two Napoleons, and two Whitworth rifles which belonged to Captain W.B. Hurt's battery were located next to the Chambersburg Pike. Then continuing the line north of the pike: Captain Joseph McGraw's four Napoleons, Lieutenant William E. Zimmerman's three three-inch rifles (the fourth having been disabled as it came into position). It is also worth noting that in his official report Lieutenant Colonel John Garnett, who was in command of Major General Heth's artillery battalion, sent Pegram Captain V. Maurin's battery (Garnett states six rifled pieces but Maurin's battery only contained three). For battery specifics see Busey and Martin, *Regimental Strengths and Losses*, 178, 192-193.

[2] *New York at Gettysburg,* 1002. A Federal officer commented that his unit's position became "untenable on account of heavy artillery fire from the north and west..."

[3] Hartwig, "The Defense of McPherson's Ridge," 19.

[4] *Ibid*, 20.

South of the 149th the guns of Reynold's Battery L, 1st New York Light Artillery (now under the command of Lieutenant George Breck after Captain Reynolds received a nasty injury to his eye) continued to duel with Carter's guns on Oak Hill.[5] Captain James H. Cooper's Battery B, 1st Pennsylvania, was assisting.[6] Cooper's rifles were in an oat field approximately two hundred yards behind Reynolds' cannon.[7]

Soon the fire from Oak Hill subsided slightly, and Wainwright ordered Cooper's guns to a new position on Seminary Ridge.[8] Reynolds' and Cooper's cannon apparently inflicted some damage on the Confederate pieces; Carter reported "having 4 men killed outright and 7 more or less severely wounded."[9]

Responding to a request from General Wadsworth, Colonel Wainwright then adjusted Battery L's guns. Dividing the unit, Wainwright sent Lieutenant Wilber's section to the orchard just south of the McPherson buildings (these structures helped shield them from Carter's fire), while Breck's remaining sections moved to their left rear (southwest), joining a line of infantry (part of Colonel Chapman Biddle's brigade) on the crest of the hill.[10]

With their artillery concentrating a heavy fire on the Union positions, the Confederate infantry began their advance. Brigadier General Junius Daniel's North Carolina regiments moved south from Oak Hill. Three of his regiments supported Brigadier General Alfred Iverson's brigade, while Daniel led the 2nd North Carolina Battalion and the 45th North Carolina toward the Chambersburg Pike – defended by the 149th Pennsylvania. To counter this threat, Colonel Stone ordered his 143rd Pennsylvania to change front to the north, supporting the 149th.[11]

---

[5] *O.R.*, Vol. 27, Part I, 362; Hassler, *Crisis at the Crossroads*, 91.

[6] *O.R.*, Vol. 27, Part I, 364.

[7] *Ibid.,* 355-356 and 364; Hartsig, "The Defense of McPherson Ridge", map on page 18; Busey and Martin, *Regimental Strengths and Losses*, 30.

[8] *O.R.*, Vol. 27, Part I, 356; *Wainwright Journal*, missing pages; Wainwright to Bachelder, May 12, 1878, *Bachelder Papers*.

[9] *O.R.*, Vol. 27, Part II, 602; Hassler, *Crisis at the Crossroads*, 91.

[10] *O.R.*, Vol. 27, Part I, 362. The battery changed front to face west.

[11] Hartsig, "The Defense of McPherson Ridge," 20.

Daniel's advance drew the attention of some Federal artillery. Three of Cooper's guns (in their new position on Seminary Ridge) "fired through the interval" between Cutler's and Stone's brigades at the butternut and gray troops.[12] The 149th Pennsylvania delivered "a staggering volley" into Daniel's ranks and then "charged, driving them back over the fence in utter confusion." After the Confederates retired, the 149th quickly moved back to the railroad cut.[13]

Upon returning "to the cut, [they] found that the enemy had planted a battery which perfectly enfiladed [their position] and made it untenable..."[14] Most likely this fire came from Captain T.A. Brander's battery, whose four guns had earlier crossed Willoughby Run to support Brigadier General Joseph Davis' Brigade. Assuming these guns advanced from their elevated position north of the Bender farm to a point along the cut, they could have fired due east along the railroad bed. From here they could "open upon [the Federal] infantry, which they did in handsome style..."[15]

At approximately 2:30 p.m., General Heth's division began crossing Willoughby Run in force; coinciding with this advance, the Confederate artillery initiated a cannonade hoping to soften the

---

[12] *O.R.*, Vol. 27, Part I, 248; Hassler, *Crisis at the Crossroads*, 103; In his official report Captain Cooper stated that only "three guns were engaged, one axle having broken from recoil at the first few shots." *O.R.*, Vol. 27, Part I, 365; Lieutenant James Stewart's Battery B, 4th U.S. Artillery, which was in position on each side of the railroad cut on Seminary Ridge, might have also participated in this fire. See, Hartsig, "The Defense of McPherson Ridge," 20;

[13] Hartsig, "The Defense of McPherson Ridge," 20.

[14] *O.R.*, Vol. 27, Part I, 330.

[15] *Ibid.*, Part II, 678. Additional evidence suggesting this was Brander's battery is in Abner Doubleday, *Chancellorsville and Gettysburg* (New York, 1882), 144. "Daniels brought artillery to enfilade the cut"; In the official reports, Brander's battery was stated to have been "about 500 yards from the enemy's batteries." The enemy battery could have been Lieutenant Wilber's section of Battery L, 1st New York, in position near the buildings on McPherson Ridge; Also see Hartsig, "The Defense of McPherson Ridge", 21.

Federal defenses on McPherson Ridge.[16] Even with this support Heth's lead brigades, under the command of Brigadier General James Pettigrew and Colonel John Brockenbrough, still encountered heavy resistance.

Watching the Confederate assault from Lieutenant Breck's position (Battery L, 1st New York), Colonel Wainwright became concerned. "There was not a shadow of a chance of our holding this ridge," he later wrote, "...[because] they outflanked us at least half a mile."[17] Breck's pieces sent shells into the Confederate lines coming over the ridge but only managed to get off about six rounds before their efforts were "much interfered with by our own infantry moving in front of his pieces." Wainwright then ordered the battery to retire.[18]

Before long the outnumbered Union troops were forced back to the Seminary, where Major General Abner Doubleday scrambled to establish a new defensive line. The artillery would provide a major part of his firepower. Lieutenant Stewart's Battery B, 4th U.S. Artillery, was north of the Chambersburg Pike – with three Napoleons on each side of the railroad cut. South of the pike and closer to the Seminary were Captain Greenleaf Stevens' (5th Maine) six Napoleons, Captain Cooper's three rifles, and then Lieutenant Wilber's section from Battery L, 1st New York. These pieces were concentrated "in so small a space that they were hardly 5 yards apart." Lieutenant Breck's remaining sections of Battery L were south of the Seminary, "behind a strong stone wall on the Seminary crest."[19]

---

[16] Hartsig, "The Defense of McPherson Ridge," 20 and 23; Hassler, *Crisis at the Crossroads*, 110; *O.R.*, Vol. 27, Part II, 678; *Bachelder Map*.

[17] *Diary of Battle*, 235. The Confederate lines overlapped the Union left.

[18] *O.R.*, Vol. 27, Part I, 356 and 362.

[19] *Ibid.*, 356. These dispositions were from Wainwright's official report. Also see, *O.R.*, Vol. 27, Part I, 362-363, and Hassler, *Crisis at the Crossroads*, 120. As for the batteries being only five yards apart, regulations called for fourteen-yard intervals between guns. Coggins, *Arms and Equipment of the Civil War*, 71; Pfanz, *Gettysburg: The Second Day*, 127. The disadvantages caused by crowding artillery pieces were that they were harder to service and presented an easier target for enemy counter battery fire.

These field pieces helped make the ridge a formidable barrier to a frontal assault. The main weakness, however, was the potential for being flanked from the south – the Federals lacked the manpower to extend their line.

After gaining McPherson Ridge, the Confederates paused to regroup before pressing on toward the next Union stronghold. Major General William Dorsey Pender's fresh division moved to the front, replacing Heth's exhausted troops.[20] The Confederate batteries on Herr Ridge were also active during this period. Pegram's and McIntosh's battalions followed their infantry, hoping to provide support for the next assault. Apparently, the only guns to reach McPherson Ridge in time to engage the enemy were two batteries under McIntosh's command.[21]

Once General Pender's brigades were aligned, the attack was renewed. Union infantry and artillery opened a murderous fire on their lines, but the Confederates pressed on. The cannon were most effective now that they could fire rounds of canister. One Southern officer commented that the Federal "artillery… maintained a constant and most galling fire upon us..."[22] Brigadier General Alfred Scales' brigade received the heaviest fire – Battery B, 4th U.S. Artillery's guns unleashed canister into their flank.

Three of Battery B's cannon were located between the Chambersburg Pike and the railroad cut. Loaded with double canister, the crews waited for Scales' troops to reach "what we thought about three hundred yards." At that point Lieutenant Davison, in command of the three guns on this side of the pike, shouted the order to fire. The blasts were devastating and the Confederate lines wavered and then fell back.[23]

---

[20] Coddington, *The Gettysburg Campaign*, 293.

[21] *O.R.*, Vol. 27, Part II, 675. McIntosh in his official report discusses enfilading the Union line from Herr Ridge and then he states, "Previous to this time, I had advanced two of my batteries to the intervening hollow, and followed close upon the enemy as he left the hills."

[22] *Ibid.*, 661.

[23] Ken Brady and Florence Freeland (eds.), *The Gettysburg Papers* (Dayton, Ohio, 1978), 369. This is a personal account of Brevet Major James Stewart's experience at Gettysburg, reprinted from *Sketches of War History* (Ohio), Volume IV. Stewart's description of the events is a bit confusing.

Lieutenant Stewart was with the guns during the action. He told Davison to employ the same technique if the Confederates approached again and then rode over to check on his other three pieces (located across the railroad cut and a little further to the west). He quickly prepared these guns for the expected attack, changing front slightly (southwest rather than west), so as to fire into the Confederate's flank. "I then had canister brought up," remembered Stewart, "and the guns were double-shotted and waited for them."[24]

Before long Scales' brigade again advanced toward the northern end of Seminary Ridge. As they marched, the three pieces north of the railroad cut remained undetected. Stewart waited until Scales' left flank was exposed and then ordered the double canister unleashed. "Here the brigade encountered a most terrific fire of grape and shell on our flank, and grape and musketry in our front," Scales later wrote. "Every discharge made sad havoc in our line, but still we pressed on at a double-quick until

---

At first it is difficult to determine for sure which part of the battle he is talking about. The main problem is that after the description of the events he states, "it was about 2:00 o'clock when the enemy made another assault...for over two hours the fighting was terrific." (page 370) Based on this statement one would believe that his initial description dealt with the repulse of Daniel's brigade. However, there are some clues that I believe establish the events that Stewart described as taking place during the defense of Seminary Ridge. First, he states after the description of events listed, "The troops we had been fighting up to this time were Heth's division of A.P. Hill's corps." (page 370) Daniel's Brigade was part of Rodes' division, of Ewell's Corps. Second, he wrote, "the enemy were not aware of us being on the other side of the railroad cut, for as they advanced they exposed their left flank."(page 369, Italics added) Since the half-battery he was commanding was on the same side (north) of the railroad cut as Daniel's brigade, but on the opposite side from Pender's brigades, this would also suggest he was describing the repulse of forces attacking Seminary Ridge.

Finally, Lieutenant Colonel Rufus Dawes who was the commander of the regiment supporting Battery B, described the events similar to Stewart. His description was clearly about the attack of Pender's division however. Rufus Dawes, *Service With the Sixth Wisconsin Volunteers*, (1890, reprint 1962), 175.

[24] *The Gettysburg Papers,* 369-370.

we reached" a position about seventy-five yards from the Seminary.[25]

Lieutenant Davison then "swung his guns around on the Cashtown Pike, so as to enfilade the whole line."[26] Scales' brigade melted away, "only a squad here and there marked the place where regiments had rested." Only one field officer in the brigade escaped unharmed, and nearly 500 of the men were casualties.[27] The fire was "undoubtedly the most destructive field artillery fire by a battery of the entire battle – perhaps of the war!"[28]

As devastating as the fire from the Federal muskets and artillery was, it was not enough to hold off the Confederate assault. With the rebel lines overlapping the Union left, and seeing the obvious retreat of the Eleventh Corps in their rear, the inevitable command for the First Corps to retire was given.

These orders confused Colonel Wainwright; earlier he received a message from General Howard that S*eminary* Hill was to be held at all costs. Unfortunately, the aid who delivered this message spoke with broken English, and what he was trying to convey was that *Cemetery* Hill should be held at all costs.[29] When Wainwright saw the infantry units leaving Seminary Ridge he was astonished. Believing the ridge should not be abandoned, he initially refused to allow his pieces to head for the rear. Finally, with the support leaving, he reluctantly ordered his guns to be hitched to the limbers. It was almost too late.

Once hitched, the guns did not immediately rush to safety; Wainwright wanted to conduct an orderly retreat. Concerned that his command could cause a panic if they roared past the infantry clogging the roads, he instructed the limbers to move at a walk down the slope. They continued at a deliberate pace until the road was relatively clear of soldiers (many of the infantrymen preferred to retire down the more protected railroad cut). Wainwright then yelled, "Trot!" and soon "Gallop!" His pieces rumbled down the

---

[25] *O.R.*, Vol. 27, Part II, 670.

[26] *Ibid*, Part I, 357.

[27] *Ibid,* Part II, 670.

[28] Hassler, *Crisis at the Crossroads*, 121.

[29] *Diary of Battle*, 235 - 236.

road three abreast.[30] This action showed the skill and discipline of Wainwright's artillerymen, as they resisted the temptation to rush for safety.[31]

Not all of Wainwright's guns escaped (and if not for the bold rearguard action of the Iron Brigade, more guns might have been lost). As Battery L, 1st New York's last section was leaving the field a horse was shot. Three others fell in rapid succession. With the Southern troops "rapidly advancing, the piece disabled was obliged to be abandoned, and fell into the enemy's possession."[32]

Lieutenant Stewart also had trouble extricating his cannon – the rough ground around the railroad cut made escape difficult. At first he moved the three pieces through the woods to his rear but then decided to cross the cut and make for the road. The first two guns made it over the large rocks, but the pintle hook on the third limber broke while negotiating the rough terrain. Stewart ordered the first two pieces to drop tail and fire at the approaching enemy, while the men in the cut scrambled to attach the gun's prolong rope to the limber. This done, they managed to get the piece up the slope to the road (probably with the help of some members of the 14th Brooklyn), and the three cannon headed for the rear.[33]

The Federal retreat offered what one Confederate officer termed, "an artillerist's heaven." That is to "throw shells and canister into [the enemy's] disorganized and fleeing masses."[34] The Southern gunners advanced several pieces to Seminary Ridge in

---

[30] *Ibid*, 235 - 236.

[31] Historian Harry W. Pfanz noted, "...the well-drilled First Corps batteries did themselves proud – they held to a walk even on the down- slope when the temptation to increase their gait must have been exceedingly strong." Pfanz, *Gettysburg: Culp's Hill and Cemetery Hil*, 50.

[32] *O.R.*, Vol. 27, Part I, 363; Sean McAdoo, *Rochester's Elite: A Brief History of Company L, 1st Regiment New York Light Artillery*, 17; *Diary of Battle*, 237. The Confederates consequently used this captured gun until the following spring when it was recaptured by the Federals at Spotsylvania Court House. See James C. Hazlett, "The 3-Inch Ordinance Rifle" from *Civil War Times Illustrated*, Vol. VII. Number 8, 35.

[33] *The Gettysburg Papers*, 371-372; C.V. Tevis, *The History of the Fighting Fourteenth* [Brooklyn], New York, 1911), 86.

[34] Confederate artillery commander E.P. Alexander, *Battles and Leaders*, Vol. III, 360.

time to fire on the Federals scrambling through town. One of these shells destroyed a caisson in Stewart's battery.[35] Another of Stewart's caissons became disabled when a wheel broke. "The artillerymen coolly stopped and in the face of the fierce fire jacked up the caisson and replaced the wheel..."[36] With this the men and equipment continued on toward Cemetery Hill.

With the fields north and west of town in Confederate hands, the batteries in the First and Eleventh Corps quickly made their way through the village streets to Cemetery Hill. Here the Federals again tried to establish a defensive line.

---

[35] Pfanz, *Culp's Hill and Cemetery Hill*, 50 - 51; Wise, *The Long Arm of Lee*, 622.
[36] *History of the Fighting Fourteenth*, 87.

Chapter Eight

Cemetery Hill

At about 4:30 p.m., Major General Winfield Scott Hancock arrived at Gettysburg with orders from General Meade to assume command of the field. With the remnants of the Eleventh Corps streaming out of the village, Hancock and the other officers attempted to rally them and form a defensive line. General Oliver Howard had previously placed Colonel Orland Smith's brigade of infantry and Captain Michael Wiedrich's 1st New York Light Artillery, Battery I, in reserve on Cemetery Hill. These fresh units were a welcome sight to the retreating troops.[1]

When Colonel Wainwright arrived on Cemetery Hill, just ahead of his batteries, he met General Howard. Howard asked Wainwright to assume command of both the First and Eleventh Corps batteries and to prepare the guns for a defensive stand. The general then put his chief of artillery, Major Osborn, at Wainwright's disposal. He then gave both men a general description of the dispositions he desired.[2]

Wainwright and Osborn agreed to divide the task of positioning the batteries; Wainwright would be responsible for the batteries on East Cemetery Hill, while Osborn oversaw the dispositions on the west side of the Baltimore Pike.[3] Wainwright arranged the guns "so as to command the approaches from the town from the northwest in case the enemy should attempt to follow us

---

[1] Hassler, *Crisis at the Crossroads*, 132, 136.

[2] Pfanz, *Culp's Hill and Cemetery Hill*, 54.

[3] *Ibid.*, 54.

through town."[4] He adjusted Weidrich's guns slightly and then detached one of his sections to Osborn's left. Cooper's three rifled pieces were positioned to the right of Weidrich's remaining sections. Reynolds' five ordnance rifles were a little to Cooper's right and rear. These guns could cover the open approaches east of town.[5]

Wainwright then placed Lieutenant Stewart's Napoleons behind, and perpendicular to, this row of rifles. From this position in front of the cemetery gatehouse, these four guns covered the Baltimore Pike. Any troops exiting the village on this road would face rounds of double canister from these brass tubes.

Captain Stevens' 5th Maine temporarily halted beside the cemetery gatehouse and awaited orders. A general whom Stevens did not immediately recognize called out asking for the commander of the "brass battery." Reporting to the general, Stevens was instructed to move his pieces off to the right. The officer pointed toward a knoll to the southeast and slightly ahead of Cemetery Hill and told the captain to place his guns there so they could cover the approaches to the elevation. Stevens asked by whose authority this was to be executed. "General Hancock's," bellowed the general. The wide eyed Stevens turned to his command and shouted, "Fifth battery, forward!"[6]

Hancock's placement of the 5th Maine was of great tactical importance. Although the Confederates did not attack that evening, they did on the next (July 2). From this location Stevens' guns poured a devastating fire into the Confederate's exposed left flank. They covered the "dead space" immediately in front of Cemetery Hill – where the guns on the crest could not depress their elevation enough to hit.[7]

Wainwright's gun positioning on Cemetery Hill was also tactically sound. Using his rifled pieces to cover the more distant approaches, Wainwright hoped to break up the formations before they reached the Union lines. Then by placing Stewart's

---

[4] *O.R.*, Vol. 27, Part I, 357.

[5] Pfanz, *Culp's Hill and Cemetery Hill*, 55.

[6] *Ibid*, 55 - 56.

[7] *Ibid*, 255.

Napoleons to protect the approaches from the village, he was maximizing the strength of each weapon.

Major Osborn also made his dispositions based on the advantages of each type of gun. He placed Dilger's and Wilkeson's Napoleons parallel to Stewart's, but behind the gatehouse. Here they could assist in defending the approach from the village. Further down the line rifled pieces covered the distant fields to the west and engaged the Confederates in counterbattery fire.[8]

In summarizing the situation on the evening of July 1st, artillery historian Van Loan Naisawald stated that the Federals had forty-one guns on Cemetery Hill waiting for an attack. Since about half of these pieces were Napoleons, "the ideal gun for this type of 'close in' defensive fighting," Ewell's infantry would have faced formidable opposition if he had attacked. "Chances are Ewell would have suffered considerably from their fire and, if his forces had faced equally determined resistance from the Blue infantry, the attack might have been beaten off." The Confederates would have sustained substantial losses from the Federal artillery fire alone.[9]

---

[8] Pfanz, *Culp's Hill and Cemetery Hill*, 57.

[9] Naisawald, *Grape and Canister*, 362.

Chapter Nine

Conclusion

At least twenty-nine batteries were engaged during the first day's battle at Gettysburg. These guns, both Confederate and Federal, were well handled and efficiently placed. The previous reorganization of artillery aided both sides, increasing efficiency and firepower. In fact, on the fields north and west of Gettysburg the Confederates had one of their best artillery showings of the entire war.[1]

The Southern guns were well placed and impressively used for supporting the infantry attacks, a difficult task. Carter's battalion probably occupied the most strategic artillery position on the field, Oak Hill. From here these batteries were able to enfilade the lines of both the First and Eleventh Corps; aside from inflicting casualties, they also influenced the dispositions of the Union troops. This was especially true west of town where the First Corps could not fully utilize the natural defensive protection of McPherson Ridge without exposing themselves to Carter's fire.

Additionally, General Jubal Early's placement of Jones' batteries just north of Rock Creek was superb. From this elevated ground his batteries could enfilade the Eleventh Corps lines and fire over the heads of the Confederate infantry as they advanced.

The Southern battery commanders demonstrated skill and aggressiveness when they advanced their guns, following up on their infantry's success. Jones' battalion sent at least one battery forward, following the retreating Federals, while Pegram's and

---

[1] Hassler, *Crisis at the Crossroads*, 101.

McIntosh's pieces also employed this tactic.[2]

Few would say this was the Federal artillerists' finest day – not because their guns were mishandled, but mainly because they were out-numbered and deployed on exposed ground. The Confederates were able to bring more infantry and artillery to the field of battle compared to the Federals. It is important to remember, however, that according to the goals set by both Buford and Reynolds earlier that morning – to fight a delaying action west and north of town until the remainder of the army could occupy the strategic hills south of the village – the Union forces were successful. A substantial part of this success can be attributed to their batteries. These guns were bravely served and provided both impressive counterbattery fire and infantry support.

Even with the limited number of guns available to the Federals on the first, there were signs that the new brigade system was beneficial. This was especially apparent during the final hours of the battle when Wainwright and Osborn established an impressive defensive wall of artillery.[3] Here these officers had both the latitude and guns available to form a coordinated line – utilizing the individual strengths of the pieces at their disposal.

There are several Federal battery commanders deserving of special mention. Captain Hubert Dilger's innovative and aggressive tactics allowed his battery, and that of Lieutenant Wheeler, to maximize their firepower when dueling with a larger number of Southern guns. Additionally, Lieutenants Stewart and Davison displayed impressive skills when handling their cannon on Seminary Hill. These two maximized the effect of their discharges by waiting for the Confederate infantry to come within the most effective range and at the most effective angle before they unleashed their rounds of canister. Captain Hall's tactics in using the terrain to its fullest advantage on McPherson Ridge (as a shield for his limbers and in extricating his guns) were also impressive.

Finally, Colonel Charles Wainwright's leadership during the

---

[2] Carrington's battery was ordered forward, but it did not have the opportunity to come into battery as the Federal retreat was so rapid. Hassler, *Crisis at the Crossroads*, 81.

[3] Naisawald, *Grape and Canister*, 444.

first days' conflict is also worthy of praise. Although limited in his latitude for gun placement, he demonstrated a good understanding of the terrain and an appreciation for the strengths and weaknesses of the rifled and smoothbore cannon. He also appreciated the fact that he needed to cooperate with General Wadsworth, while at the same time trying to effectively place his guns.

Several Confederate gunners also deserve recognition. Professor Hassler, in his summary of the first day's fighting wrote, "special laurels must be given to the Confederate artillery battalions of Jones and Carter, which played leading roles in helping the Southern infantry finally to throw back Schurz and Doubleday."[4] Added to this list should be the young Pegram. He returned to duty right before the battle and handled his guns superbly. Additionally, his mere presence positively affected both his subordinates and his superiors.

The previous experience and skill that these artillerists (both Union and Confederate) brought to the field, coupled with the reorganization of artillery, enabled both sides to handle their field pieces in an impressive manner. These men showed their skill, professionalism and bravery. The artillerists that fought at Gettysburg would have rivaled or surpassed any cannoneers in the world at that time.

---

[4] Hassler, *Crisis at the Crossroads*, 154.

Appendix One

The Mechanics of Loading and Firing a Civil War Artillery Piece

Loading and firing light artillery pieces required a crew of eight men. Four of the men actually serviced the piece, two remained with the limber and handled the ammunition, one transported the ammunition from the limber to the cannon, and one gunner oversaw the sighting and firing. Additionally, the Federals had a chief of piece who was in command of the gun.

The following is how an actual artillery officer described the duties of the gun crew in loading and firing a field piece. The officer was Lieutenant Henry Vaughn, and he served with the First New York Independent Artillery Battery, which saw heavy action at Gettysburg.

> **No. 1**...He has the hardest place, which is to ram home the charges. After each firing he sponges out the gun & rams home the cartridge and shell which is put into the muzzle by
> **No. 2**...His duty is merely to place the charge in the muzzle powder bag first & shell afterwards, & then when No. 1 is tired he alternates with him.
> **No. 3**...He has a thumbstall made of leather and a pricking wire & puts his thumb...over the vent of the gun when No. 1 is wiping out with a sponge so that all fire may be extinguished before the new load is put in...
> **No. 4**...drops into the vent a priming tube which is a tube filled with powder having at the top another tube soldered [sic] at right

angles with a rough wire enclosed in percussion powder which is a rough wire pulled out by No. 4 at the command "Fire." To facilitate the pulling out of the wire a lanyard (a small rope) with a hook attached is hooked into the ring made by the end of the rough wire & then [No. 4] standing well outside of the wheel of the cannon, jerks the lanyard briskly pulling the wire out of the percussion, igniting it & setting fire to the powder in the tube which runs down into the cartridge & away goes the shell with a crack & jar that shake the earth under our feet but that's the "music we like."

No. 5...duty of running to the ammunition chest of the gun and bringing the charges as wanted to No 2 who inserts them.

No. 6...His work is to cut out fuses & stick them into the shells & to send fuses & different kinds of ammunition as it may be ordered by the gunner...

No. 7...His duty is to assist No 6 by holding the shells while No 6 prepares them & to alternate with No 5 when he gets tired.

No 8...his duty is to stay with the caisson in the rear & keep the gun ammunition supplied when in action...

The gunner...points and aims the gun at command of "Load," & "Ready," & "Fire."

Appendix Two

Glossary of Terms

Please note: This is **not** intended to be a manual of exact definitions, but rather a general tool for those who are unfamiliar with the terms used in this book.

ammunition chest A wooden box that contained ammunition for the cannon. Each limber had one chest, and the caissons carried two. A chest loaded for a 12-pound Napoleon carried thirty-two rounds of ordnance and weighed about 500 lbs. (twelve solid shot, twelve case shot, four shells, and four rounds of canister). (See Ripley, *Artillery and Ammunition of the Civil War,* 196).

artillery battalion An organization of Confederate artillery batteries assigned to divisions or corps reserve. Battalions that served in the Army of Northern Virginia at Gettysburg are listed below.

First Corps – contained five artillery battalions, attached to:
- McLaws' Division – battalion of four batteries (Col. Cabell).
- Pickett's Division – battalion of four batteries (Maj. Dearing).
- Hood's Division – battalion of four batteries (Maj. Henry).
- Artillery Reserve (Corps)
  - Alexander's battalion – battalion of six batteries (Col. Alexander).
  - Washington Artillery – battalion of four batteries (Maj. Eshle man).

Second Corps – contained five artillery battalions, attached to:
- Early's Division – battalion of four batteries (Lt. Col. Jones).
- Johnson's Division – battalion of four batteries (Maj. Latimer).
- Rodes' Division – battalion of four batteries (Lt. Col. Carter).
- Artillery Reserve (Colonel Brown).
  - 1st Virginia Battalion – battalion of five batteries (Capt. Dance).
  - Nelson's Battalion – battalion of three batteries (Lt. Col. Nelson).

Third Corps – contained five artillery battalions, attached to:
- – Anderson's Division – battalion of three batteries (Maj. Lane).
- – Heth's Division – battalion of four batteries (Lt. Col. Garnett).
- – Pender's Division – battalion of four batteries (Maj. Poague).
- – Artillery Reserve (Colonel Walker).
  - – McIntosh's Battalion – battalion of four batteries (Maj. McIntosh).
  - – Pegram's Battalion – battalion of five batteries (Maj. Pehram).

Cavalry Corps – contained two artillery battalions, attached to:
- – Stuart Horse Artillery – battalion of six batteries (Maj. Beckham).
- – Imboden's Command – battalion of four batteries (Brig. Gen. Imbo den).

artillery brigade An organization of Union artillery batteries attached to a corps or assigned to the artillery reserve. Artillery brigades that served in the Army of the Potomac at Gettysburg are listed below.

First Corps – artillery brigade of five batteries (Col. Wainwright).
Second Corps – artillery brigade of five batteries (Capt. Hazard).
Third Corps – artillery brigade of five batteries (Capt. Randolph).
Fifth Corps – artillery brigade of five batteries. (Capt. Martin).
Sixth Corps – artillery brigade of eight batteries. (Col. Tompkins).
Eleventh Corps – artillery brigade of five batteries. (Maj. Osborn).
Twelfth Corps – artillery brigade of four batteries. (Lt. Muhlenberg).
Cavalry Corps
- – First Brigade – artillery brigade of five batteries. (Capt. Robertson).
- – Second Brigade – artillery brigade of four batteries. (Capt. Tidball).

Artillery Reserve
- – First Regular Brigade – four batteries. (Capt. Ransom).
- – First Volunteer Brigade – four batteries (Lt. Col. McGilvery).
- – Second Volunteer Brigade – four batteries. (Capt. Taft).
- – Third Volunteer Brigade – four batteries. (Capt. Huntington).
- – Fourth Volunteer Brigade – five batteries. (Capt. Fitzhugh).

battery The basic command unit for artillery pieces. Typically Confederate batteries at Gettysburg contained four guns, while the Federal batteries contained six pieces.

caisson A two wheeled ammunition carrier for the battery. Typically each battery contained one caisson for each artillery piece. Aside from carrying two ammunition chests, there was also a spare wheel attached to the caisson that fit the wheels of the limbers and gun carriages. The caisson was pulled by a team of six horses and was hitched to an artillery limber. (See Hazlett, Olmstead and Parks, *Field Artillery Weapons of the Civil War,* 219).

canister: Ordnance fired by artillery at close range. These tin cans filled

with small iron or lead balls were packed in sawdust, and turned the cannon into a giant shotgun. It was the most effective type of ordnance fired by cannon, but could only be used at close range (under 400 yards). (See Naisawald, *Grape and Canister,* 539)
case shot (or Shrapnel) Projectiles fired from a field piece with a powder charge inside that was ignited by a fuse. Ideally it would explode fifty to seventy-five yards out in front of the enemy, and about twenty feet up. The explosion caused marble sized balls to spray into the enemy position. It was primarily an anti-personnel weapon. (See Naisawald, *Grape and Canister,* 539, and Ripley, *Artillery and Ammunition of the Civil War*).
horses A battery required the service of dozens of horses. Field pieces and caissons were attached to limbers, each with a team of six horses (Confederate batteries were often forced into using four-horse teams to pull the artillery, especially later in the war). Additionally, the officers and some of the sergeants (buglers and color bearers) were mounted. Horses or mules might have pulled the forge and supply wagons.
horses – wheel pair The two horses on the limber team closest to the limber.
horses – swing pair The two horses in the middle of the six-horse team.
horses – lead pair The two horses at the front of the team.
limber A two wheeled carriage that carried ammunition for the gun and towed either a cannon or a caisson. It was pulled by a team of six horses and carried three seated crewmembers on the limber chest.
limber chest The ammunition chest on the limber. This chest supplied the individual guns with ordnance during battle.
Napoleon Officially named the "Light 12-pounder gun," it became the most popular smoothbore cannon during the conflict. General George McClellan's chief of artillery even recommended that Napoleons should be made the exclusive smoothbore field piece of the Army of the Potomac. The 12-pounder got its nickname from Prince Charles Louis Napoleon Bonaparte, *the* Napoleon's nephew, because he was the driving force behind its manufacture and use in the French army. (See Hazlett, Olmstead, and Parks, *Field Artillery Weapons of the Civil War*).
ordnance: A typical limber chest for a light artillery piece contained four types of rounds: solid shot, case, exploding shells and canister.
Ordnance rifle (also three-inch rifle) Common rifled cannon used by both armies during the war. It was very reliable and accurate.
prolonge A rope that could be attached to the cannon and the limber, allowing the guns to be towed to the rear slowly while the crews loaded and fired at the enemy.
section Two artillery pieces made up a section.

shells (exploding shells) Fired from a field piece, this thick walled but powder filled projectile was intended to explode in front of an enemy position and break into pieces. There was a major psychological factor associated with this weapon, mainly because of its noise when it exploded. (See Naisawald, *Grape and Canister,* 538).
shot (solid shot) A solid metal projectile fired from a field piece. It did not explode or fragment on impact, but its inertia made it dangerous because it often continued on through ranks of men inflicted heavy casualties .
spacing:
between each piece (wheel to wheel) – fourteen yards.
between cannon and lead limber team – six yards.
tools:
friction primer An ignition device made of two brass tubes at right angles and soldered together. The vertical tube was filled with powder and inserted into the vent of the cannon. When the lanyard was jerked the friction created by the top (horizontal) tube caused the friction powder to ignite. This fire caused the powder inside of the cannon to explode. (See Naisawald, *Grape and Canister,* 549-550).
handspike A wooden handle that fit into two rings (pointing rings) on the back (trail) of the field piece. The Number Three man would use the trail handspike for leverage when adjusting the cannon according to the Gunners directions, "Trail Right" or "Trail Left." (See Coggins, *Arms and Equipment of the Civil War*, 70).
lanyard A rope with a hook on the end that attaches to the friction primer. The Number Four jerks the lanyard rope that ignites the friction primer.
sponge and rammer staff Tool used by the Number One man to first sponge out the barrel (to extinguish any remaining embers from the previous discharge) and then using the rammer, "ram home" the charge (forcing it down the tube to the breech – under the vent tube).
thumb stall Leather protective cover for the Number Three man's thumb, which he placed over the vent while the piece was being loaded.
vent pick Tool inserted down the vent tube to prick a hole in the powder bag. This allowed the lower end of the friction primer to contact the powder.

Bibliography

Alexander, E.P. *Military Memoirs of a Confederate*. New York: Charles Scribner's Sons, 1907.

_____ "The Great Charge and Artillery Fighting at Gettysburg" *Battles and Leaders of the Civil War*, Vol. III.

Brady, Ken and Florence Freeland. eds. *The Gettysburg Papers* Dayton, Ohio: Morningside Press, 1978.

Boatner, Mark M. III. *The Civil War Dictionary*. New York: David McKay Co., 1959.

Busey, John W. and David Martin, *Regimental Strengths and Losses at Gettysburg*. Hightstown, N.J.: Longstreet House, 1982.

Chandler, David. ed., *The Military Maxims of Napoleon.* Stack pole Books edition, 1994.

Coddington, Edwin B. *The Gettysburg Campaign: A Study in Command.* New York: Scribners, 1968.

Coggins, Jack. *Arms and Equipment of the Civil War*. New York: Fairfax Press, 1962.

Crumb, Herb. ed. *The Eleventh Corps Artillery at Gettysburg: The Papers of Major Thomas Ward Osborn, Chief of Artillery*. Hamilton, New York: Edmonston Press, 1991.

Crumb, Herb, and Katherine Dhalle, eds. *No Middle Ground: Tho mas Ward Osborn's Letters from the Field (1862 – 1864).* Hamilton, New York: Edmonston Press, 1993.

Dawes, Rufus. *Service With the Sixth Wisconsin Volunteers*. Mari etta, Ohio: E.R. Alderman and Sons, 1890.

Doubleday, Abner. *Chancellorsville and Gettysburg.* New York:Charles Scribner's Sons, 1882.

Downey, Fairfax. *Guns at Gettysburg.* New York: David McKay Co., 1958.

Eisenhower, John S.D. *So Far From God: The U.S. War with Mexico 1846 – 1848.* New York: Doubleday, 1989.

Esposito, Vincent J. ed. *West Point Atlas of American Wars,* 2 vols. New York: Frederick Praeger Publishers, 1959.

Fox, William F. *New York at Gettysburg*. Albany: J.B. Lyon, Co., 1900.

Freeman, Douglas Southall. *Lee's Lieutenants: A Study in Com mand*, 3 vols. New York: Charles Scribner's Sons, 1943.

Furgurson, Ernest B. *Chancellorsville 1863: The Souls of the Brave.* New York: Alfred A Knopf, 1992.

Griffith, Paddy. *Battle Tactics of the Civil War.* New Haven:Yale University Press, 1987.

Hall, James A. Letters to Colonel John B. Bachelder, February 27, 1867 and December 29, 1869. These letters are contained in *The Bachalder Papers*.

Hartwig, D. Scott. "The Defense of McPherson Ridge" *The Gettysburg Magazine,* Issue Number One, July, 1989.

_____ "The 11th Army Corps on July 1, 1863" *The Gettysburg Magazine*, Issue Number two, January 1, 1990.

Hassler, Warren W. Jr. *Crisis at the Crossroads: The First Day at Gettysburg.* University of Alabama Press, 1970.

Hazlett, James C., Edwin Olmstead, and M. Hume Parks. *Field Artillery Weapons of the Civil War.* Newark, N.J.: University of Delaware Press, 1988.

Hazlett, James C. "The 3-Inch Ordinance Rifle" *Civil War Times Illustrated*, Vol. VII. Number 8.

Herdegen, Lance and William J.K. Beaudot. *In the Bloody Rail road Cut at Gettysburg.* Dayton, Ohio: Morningside Press, 1990.

Hunt, Henry. "The First Day at Gettysburg", from *Battles and Leaders of the Civil War*, Vol. III.

Johnson, Robert Underwood and Clarence Clough Buel. eds. *Bat tles and Leaders of the Civil War.* New York: The Century Company, 1887.

Kepf, Kenneth M. "Dilger's Battery at Gettysburg," *The Gettys burg Magazine* January 1, 1991, Issue Number Four.

Keyser, Carl. "Leatherbreeches at Chancellorsville" *Civil War Times Illustrated*, August, 1975.

Kurtz, Henry. "Dilger's Battery," *Civil War Times Illustrated*, Vol. 1, Number 7, 39.

Ladd, David and Audrey Ladd, eds. *The Bachalder Papers.* Day ton, Ohio: Morningside Press, 1994.

Longacre, Edward G. *The Man Behind the Guns: A Biography of Henry Jackson Hunt, Chief of Artillery, Army of the Potomac.*

New York: A.S. Barnes and Co., 1977.
_____ *The Cavalry at Gettysburg.* Cranbury, N.J.: Dickenson University Press, 1986.
McAdoo, Sean. *Rochester's Elite: A Brief History of Company L, 1st Regiment New York Light Artillery,* 1996.
McWhiney, Grady and Perry Jamieson. *Attack and Die: Civil War Military Tactics and the Southern Heritage.* Tuscalusa: University of Alabama Press, 1982.
Murray, R.L. *"Hurrah for the Ould Flag!: The True Story of Captain Andrew Cowan and the First New York Independent Battery at Gettysburg.* Wolcott, NY, 1998.
Naisawald, L. Van Loan. *Grape and Canister: The Story of Field Artillery in the Army of the Potomac.* New York, 1960.
Nevins, Allan. ed. *A Diary of Battle: The Personal Journals of Colonel Charles Wainwright, 1861 – 1865.* New York: Harcourt, Brace and World, Inc., 1962.
Nosworthy, Brent. *With Musket Cannon and Sword: Battle Tactics of Napoleon and His Enemies.* New York: Sarpedon, 1996.
Pfanz, Harry W. *Gettysburg: Culp's Hill and Cemetery Hill.* Chapel Hill: University of North Carolina Press, 1993.
_____ *Gettysburg: The Second Day.* Chapel Hill: University of North Carolina Press, 1987.
Porter, Horace. *Campaigning with Grant.* New York: New Century, 1897.
Sears, Stephen W. *Chancellorsville.* Boston: Houghton Mifflin Co., 1996.
Shue, Richard S. *Morning at Willoughby Run, July 1, 1863.* Gettysburg, Pa.: Thomas Publications, 1995.
Sifakis, Stewart. *Who Was Who in the Civil War.* New York: Facts on File Publications, 1988.
Smith, A.P. *History of the Seventy-Sixth Regiment New York Volunteers.* Cortland, New York, 1867.
Tevis, C.V. *The History of the Fighting Fourteenth* [Brooklyn]. New York, 1911.
Tucker, Glenn. *High Tide at Gettysburg: The Campaign in Pennsylvania.* Bobbs-Merrill Co. Inc., 1958.
Vanderslice, John M. *Gettysburg Then and Now.* Original 1897; Morningside Press reprint edition, 1983.
The War of the Rebellion: A Compilation of the Official Records

of the Union and Confederate Armies. Washington: Government Printing Office, 1889.

Wert, Jeffrey "'Old Artillery': William Nelson Pendleton" *Civil War Times Illustrated*, June 1974.

Wise, Jennings Cropper. *The Long Arm of Lee,* or *The History of the Artillery of theArmy of Northern Virginia*. Lynchburg, Va.: J.P. Bell Co., 1915.